How to Prove The Collatz Conjecture
by Danny Karl Fleming

Introduction

On one fortunate day, while I was surfing the internet, I happened upon a site that listed

unsolved problems in mathematics. It was probably a search engine. I was extremely excited

because I have a Bachelor of Science in Mathematics which I earned from California State

University, Long Beach in 1996. So, I was bound to have a chance at "cracking the code" of

some of the unsolved problems. I found three fairly hard ones to my liking--1.Are Perfect

Perfect Numbers Infinite, 2.Goldbach Conjecture(Can any even number except 2 be

expressed as the sum of two primes?), 3.Collatz Conjecture. It is my proof of the Collatz

Conjecture I will devote my time to while writing this book

Chapter 1

The Collatz Conjecture asks if any chosen natural number (1, 2 ,3, 4, 5, 6, 7 ,8, 9, 10,

11, 12, 13, 14, ...) will reach 1 if the following rules are followed: If the chosen positive
 integer x is even, divide it by two (x/2), if it is odd, multiply it by three and add 1 (3x+1).
Examples are 01-04-02-01, 03-10-05-16-08-04-02-01, and 08-04-02-01.
I used the odd-even patterns to prove the Collatz Conjecture, which I will abbreviate CJ

hereinafter. I also used the last 2 or 3 digits of each positive integer involved in
each Collatz sequence to demonstrate the possibilities that arise after application
of the "x/2 if even" and "3x+1 if odd" rules already mentioned. When I first
started a couple of years ago, I used the last two digits to attempt to find a pattern.
The odd-even or even-odd patterns of the sequences just mentioned

are odd-even-even-odd for 01-04-02-01, odd-even-odd-even-even-

even-even-odd for 03-10-05-16-08-04-02-01, and even-even-even-odd for

08-04-02-01.

First, any even number will continue decreasing until it either reaches 1 or

an odd number. Once we reach this odd number x, we will have an even number

after the 3x+1 rule is applied. Here is why: We know 3 is always odd by definition. And

x is always odd because the 3x+1 rule is only used if x is odd. For the 3x part of 3x+1,

we have an odd number (3) times another odd number(x). We know from Book IX

Proposition 29 of Euclid's Elements that an odd number times another odd number is

always odd. Examples, using 3 and any other odd positive integer for 3x+1

are: $3x3=9$, $3x9=27$, $3x81=243$, $3x5=15$, $3x19=57$, etc. As expected, 9, 17, 243, 15,

and 57 are all odd. We know the product of 3 times odd Z^+ will always be odd

because, since odd numbers are spaced two units apart, 3 times two successive

integers will be $3x2=6$ units apart. Examples are $3x9=27$, $3x11=33$, $3x13=39$, all

separated by six units as expected (27 ,33, 39). Adding 1 to this proven odd 3x will always be an even number because any odd number plus 1 is even (3+1=4, even,

9+1=10, even, 27+1=28, even, 309+1=310, even).

We recently started using computers to verify the Collatz Conjecture. It is now verified to thousands and thousands. I found a method that would cut the work in half: The even positive integers automatically reach a number less than

themselves after application of the x/2 rule for even integers, making them coincide with

any verified odd number. This works for up to twice any verified maximum. The verified

maximum is the highest positive integer that has been verified by computers. I will

abbreviate verified maximum "VM". I am defining the verb "subcoll" to mean "an integer

subcolls if, during the process of the Collatz Conjecture, any of the results are below the

starting integer." For example, let's choose 5 as the starting integer:

5-16-08-04-...

It (the integer 5) has subcolled after the third application of the CJ because 04 is below 05.

The noun straticoll will mean "a set of integers that decrease to below their original starting

number." I am also using the verb retrocoll with a similar meaning to subcoll. The last

two or three digits are always the same for the unique even number that is generated

after the 3x+1 rule.

Here is a list of the only possible last two digit endings for any odd integer:

01--04 21--64 41--24 61--84 81--44

03--10 23—70 43—30 63—90 83—50

05--16 25--76 45--36 65--96 85--56

07--22 27--82 47--42 67--02 87--62

09--28 29--88 49--48 69--08 89--68

11--34 31--94 51--54 71--14 91--74

13--40 33--00 53--60 73--20 93--80

15--46 35--06 55--66 75--26 95--86

17--52 37--12 57--72 77--32 97--92

19--58 39--18 59--78 79--38 99--98

Here are the only possible endings for the last three digits of

any positive even integer:

x00 x10
000--000, 500 010--005, 505
100--050, 550 110--055, 555
200--100, 600 210--105, 605
300--150, 650 310--155, 655
400--200, 700 410--205, 705
500--250, 750 510--255, 755
600, 300, 800 610--305, 805
700--350, 850 710--355, 855
800--400, 900 810--405, 905
900--450, 950 910--455, 955

x02 x12
002--001, 501 012--006, 506
102--051, 551 112--056, 556
202--101, 601 212--106, 606
302--151, 651 312--156, 656
402--201, 701 412--206, 706
502--251, 751 512--256, 756
602--301, 801 612--306, 806
702--351, 851 712--356, 856
802--401, 901 812--406, 906
902--451, 951 912--456, 956

x04 x14
004--002, 502 014--007, 507
104--052, 552 114--057, 557
204--102, 602 214--107, 607
304--152, 652 314--157, 657
404--202, 702 414--207, 707
504--252, 752 514--257, 757
604--302, 802 614--307, 807
704--352, 852 714--357, 857
804--402, 902 814--407, 907
904--452, 952 914--457, 957

x06 x16
006--003, 503 016--008, 508
106--053, 553 116--058, 558
206--103, 603 216--108, 608
306--153, 653 316--158, 658
406--203, 703 416--208, 708
506--253, 753 516--258, 758
606--303, 803 616--308, 808
706--353, 853 716--358, 858
806--403, 903 816--408, 908

906--453, 953 916--458, 958

x08 x18

008--004, 504 018--009, 509

108--054, 554 118--059, 559

208--104, 604 218--109, 609

308--154, 654 318--159, 659

408--204, 704 418--209, 709

508--254, 754 518--259, 759

608--304, 804 618--309, 809

708--354, 854 718--359, 859

808--404, 904 818--409, 909

908--454, 954 918--459, 959

x20 x22

020--010, 510 022--011, 511

120--060, 560 122--061, 561

220--110, 610 222--111, 611

320--160, 660 322--161, 661

420--210, 710 422--211, 711

520--260, 760 522--261, 761

620--310, 810 622--311, 811

720--360, 860 722--361, 861

820--410, 910 822--411, 911

920--460, 960 922--461, 961

x24 x26

024--012, 512 026--013, 513

124--062, 562 126--063, 563

224--112, 612 226--113, 613

324--162, 662 326--163, 663

424--212, 712 426--213, 713

524--262, 762 526--263, 763

624--312, 812 626--313, 813

724--362, 862 726--363, 863

824--412, 912 826--413, 913

924--462, 962 926--463, 963

x28 x30

028--014, 514 030--015, 515

128--064, 564 130--065, 565

228--114, 614 230--115, 615

328--164, 664 330--165, 665

428--214, 714 430--215, 715

528--264, 764 530--265, 765

628--314, 814 630--315, 815

728--364, 864 730--365, 865

828--414, 814 830--415, 915

928--464, 964 930--465, 965

x32 x34

032--016, 516 034--017, 517
132--066, 566 134--067, 567
232--116, 616 234--117, 617
332--166, 666 334--167, 667
432--216, 716 434--217, 717
532--266, 766 534--267, 767
632--316, 816 634--317, 817

732--366, 866 734--367, 867
832--416, 916 834--417, 917
932--466, 966 934--467, 967

x36 x38

036--018, 518 038--019, 519
136--068, 568 138--069, 569

236--118, 618 238--119, 619

336--168, 668 338--169, 669

436--218, 718 438--219, 719

536--268, 768 538--269, 769

636--318, 818 638--319, 819

736--368, 868 738--369, 869

836--418, 918 838--419, 919

936--468, 968 938--469, 969

x40 x42

040--020, 520 042--021, 521

140--070, 570 142--071, 571

240--120, 620 242--121, 621

340--170, 670 342--171, 671

440--220, 720 442--221, 771

540--270, 770 542--271, 771

640--320, 820 642--321, 821

740--370, 870 742--371, 871

840--420, 920 842--421, 921

940--470, 970 942--471, 971

x44 x46

044--022, 522 046--023, 523

144--072, 572 146--073, 573

244--122, 622 246--123, 623

344--172, 672 346--173, 673

444--222, 722 446--223, 723

544--272, 772 546--273, 773

644--322, 822 646--323, 823

744--372, 872 746--373, 873

844--422, 922 846--423, 923

944--472, 972 946--473, 973

x48 x50

048--024, 524 050--025, 525

148--074, 574 150--075, 575

248--124, 624 250--125, 625

348--174, 674 350--175, 675

448--224, 724 450--225, 725

548--274, 774 550--275, 775

648--324, 824 650--325, 825

748--374, 874 750--375, 875

848--424, 924 850--425, 925

948--474, 974 950--475, 975

x52 x54

052--026, 526 054--027, 527

152--076, 576 154--077, 577

252--126, 626 254--127, 627

352--176, 676 354--177, 677

452--226, 726 454--227, 727

552--276, 776 554--277, 777

652--326, 826 654--327, 827

752--376, 876 754--377, 877

852--426, 926 854--427, 927

952--476, 976 954--477, 977

x56 x58

056--028, 528 058--029, 529

156--078, 578 158--079, 579

256--128, 628 258--129, 629

356--178, 678 358--179, 679

456--228, 728 458--229, 729

556--278, 778 558--279, 779

656--328, 828 658--329, 829

756--378, 878 758--379, 879

856--428, 928 858--429, 929

956--478, 978 958--479, 979

x60 x62

060--030, 530 062--031, 531

160--080, 580 162--081, 581

260--130, 630 262--131, 631

360--180, 680 362--181, 681

460--230, 730 462--231, 731

560--280, 780 562--281, 781

660--330, 830 662--331, 831

760--380, 880 762--381, 881

860--430, 930 862--431, 931

960--480, 980 962--481, 981

x64 x66

064--032, 532 066--033,533

164--082, 582 166--083, 583

264--132, 632 266--133, 633

364--182, 682 366--183, 683

464--232, 732 466--233, 733

564--282, 782 566--283, 783

664--332, 832 666--333, 833

764--382, 882 766--383, 883

864--432, 932 866--433, 933

964--482, 982 966--483, 983

x68 x70

068--034, 534 070--035, 535

168--084, 584 170--085, 585

268--134, 634 270--135, 635

368--184, 684 370--185, 685

468--234, 734 470--235, 735

568--284, 784 570--285, 785

668--334, 834 670--335, 835

768--384, 884 770--385, 885

868--434, 934 870--435, 935

968--484, 984 970--485, 985

x72 x74

072--036, 536 074--037, 537

172--086, 586 174--087, 587

272--136, 636 274--137, 637

372--186, 686 374--187, 687

472--236, 736 474--237, 737

572--286, 786 574--287, 787

672--336, 836 674--337, 837

772--386, 886 774--387, 887

872--436, 936 874--437, 937

972--486, 986 974--487, 987

x76 x78

076--038, 538 078--039, 539

176--088, 588 178--089, 589

276--138, 638 278--139, 639

376--188, 688 378--189, 689

476--238, 738 478--239, 739

576--288, 788 578--289, 789

676--338, 838 678--339, 839

776--388, 888 778--389, 889

876--438, 938 878--439, 939

976--488, 988 978--489, 989

x80 x82

080--040, 540 082--041, 541

180--090, 590 182--091, 591

280--140, 640 282--141, 641

380--190, 690 382-- 191, 691

480--240, 740 482--241, 741

580--290, 790 582--291, 791

680--340, 840 682--341, 841

780--390, 890 782--391, 891

880--440, 940 882--441, 941

980--490, 990 982--491, 991

x84 x86

084--042, 542 086--043, 543

184--092, 592 186--093, 593

284--142, 642 286--143, 643

384--192, 692 386--193, 693

484--242, 742 486--243, 743

584--292, 792 586--293, 793

684--342, 842 686--343, 843

784--392--892 786--393, 893

884--442, 942 886--443, 943

984--492, 992 986--493, 993

x88 x90

088--044, 544 090--045, 545

188--094, 594 190--095, 595

288--144, 644 290--145, 645

388--194, 694 390--195, 695

488--244, 744 490--245, 745

588--294, 794 590--295, 795

688--344, 844 690--345, 845

788--394, 894 790--395, 895

888--444, 944 890--445, 945

988--494, 994 990--495, 995

x92 x94

092--046, 546 094--047, 547

192--096, 596 194--097, 597

292--146, 646 294--147, 647

392--196, 696 394--197, 697

492--246, 746 494--247, 747

592--296, 796 594--297, 797

692--346, 846 694--347, 847

792--396, 896 794--397, 897

892--446, 946 894--447, 947

992--496, 996 994--497, 997

x96 x98

096--048, 548 098--049, 549

196--098, 598 198--099, 599

296--148, 648 298--149, 649

396--198, 698 398--199, 699

496--248, 748 498--249, 749

596--298, 798 598--299, 799

696--348, 848 698--349, 849

796--398, 898 798--399, 899

896--448, 948 898--449, 949

996--498, 998 998--499, 999

Here are a few 3x+1 applications for the last three digits:

101 111 121 131 141 151 161 171 181 191
x3 x3 x3 x3 x3 x3 x3 x3 x3 x3
303 333 363 393 423 453 483 513 543 573

+1=304 +1=334 +1=364 +1=394 +1=424 +1=454 +1=484 +1=514 +1=544 +1=574

201 211 221 231 241 251 261 271 281 291
x3 x3 x3 x3 x3 x3 x3 x3 x3 x3
603 633 663 693 723 753 783 813 843 873

+1=604 +1=634 +1=664 +1=694 +1=724 +1=754 +1=784 +1=814 +1=844 +1=874

301 311 321 331 341 351 361 371 381 391
x3 x3 x3 x3 x3 x3 x3 x3 x3 x3
903 933 963 993 1023 1053 1083 1113 1143 1173

+1=904 +1=934 +1=964 +1=994 +1=1024 +1=1054 +1=1084 +1=1114 +1=

1144 +1=1174

401 411 421 431 441 451 461 471 481 491
x3 x3 x3 x3 x3 x3 x3 x3 x3 x3
1203 1233 1263 1293 1323 1353 1383 1413 1443 1473

+1=1204 +1=1234 +1=1264 +1=1294 +1=1324 +1=1354 +1=1384 +1=1414 +1=1444
+1=1474

501 511 521 531 541 551 561 571 581 591
x3 x3 x3 x3 x3 x3 x3 x3 x3 x3
1503 1533 1563 1593 1623 1653 1683 1713 1743 1773

+1=1504 +1=1534 +1=1564 +1=1594 +1=1624 +1=1654 +1=1684 +1=1714 +1=1744
+1=1774

601 611 621 631 641 651 661 671 681 691
x3 x3 x3 x3 x3 x3 x3 x3 x3 x3
1803 1833 1863 1893 1923 1953 1983 2013 2043 2073

+1=1804 +1=1834 +1=1864 +1=1894 +1=1924 +1=1954 +1=1984 +1=2014 +1=2044
+1=2074

701 711 721 731 741 751 761 771 781 791
x3 x3 x3 x3 x3 x3 x3 x3 x3 x3
2103 2133 2163 2193 2223 2253 2283 2313 2343 2373

+1=2104 +1=2134 +1=2164 +1=2194 +1=2224 +1=2254 +1=2284 +1=2314 +1=2344
+1=2374

801 811 821 831 841 851 861 871 881 891
x3 x3 x3 x3 x3 x3 x3 x3 x3 x3
2403 2433 2463 2493 2523 2553 2583 2613 2643 2673

+1=2404 +1=2434 +1=2464 +1=2494 +1=2524 +1=2554 +1=2584 +1=2614 +1=2644
+1=2674

901 911 921 931 941 951 961 971 981 991
x3 x3 x3 x3 x3 x3 x3 x3 x3 x3

2703 2733 2763 2793 2853 2883 2913 2943 2973 +1=2704 +1=2734 +1=2764 +1=2794 +1=2824 +1=2854 +1=2884 +1=2914 +1=2944 +1=2974

The patterns repeat as follows: Starting with 1, every even number determines the distance

to the next odd number with that many evens and odds in the exact order (see 2^n

explanation). The distance between 1 and the next pattern equivalent to odd-even-even-

odd-even is 2^3 (#evens=3) =8 or 8+1=9. And checking with 9, the first four results are

odd-even-even-odd-even as expected. The sequences in between will follow the always

even after odd and half the time odd and half the time even (alternating odd--even or

even--odd).

These patterns are fixed and will determine the pattern of any positive integer

needed. This pattern, determined by the patterns below, will eventually coincide

with 1-4-2-1. Every time we are multiplying by 3, adding 1, dividing by 2 two times,

giving us approximately 3/4 of the starting number. This will happen after every 3

applications of the Collatz Conjecture, bringing any number below itself and therefore

reaching 1 as the ones below it already have.

Since all odd numbers are multiplied by 3 with 1 added, it is easy to explain why the above are the only possible endings for any two-digit positive integers. I have

included all possible endings for the last two digits of any positive integer

(01,03,05,07,...93,95,97,99).

Let's multiply 01 by 3, the only possible result is 03. Adding 1 to complete the 3x+1

step makes it 03+01=04.

Now try 93, 93x3=279. We know that 93 times 3 is always 279. Adding 1 because

of the 3x+1 rule for odd numbers gives us 279+1=280. As further explanation, it

doesn't matter what the other digits of any chosen odd positive integer are because

we always start at the right or one's position and move one place at a time to the left

or ten's, hundred's, thousand's positions. Examples:101x3=303, 303+1=304

(04 as predicted), 109303x3=327909, 327909+1=327910 (10 as predicted).

The patterns repeat as follows: Every positive integer will follow any positive integer

that is plus or minus any power of 2 (2^n) for n evens. Any positive integer plus any

power of 2 determines the distance to the next odd number with that many evens and odds in the exact order. I can demonstrate this with the starting integers 15 and 31

because they are a power of 2 apart ($31-15=16=2^4$). They will have the same
odd-even or even-odd pattern for the first 4 evens. Let's see:

15-46-23-70-35-106-53-160-...

31-94-47-142-71-214-107-322-...

They are both odd-even-odd-even-odd-even-odd-even.
Any chosen positive integers 2^n apart agree for n evens.
01-04-02-01-04-02-01-... will follow $2^{40}+1$ for 40 evens
03-10-05-16-08-04-02-01-04-02-01-... will follow $2^{40}+3$ for 40 evens
05-16-08-04-02-01-04-02-01-... will follow $2^{40}+5$ for 40 evens
07-22-11-34-17-52-26-13-40-20-10-05-16-08-04-02-01-04-02-01-...
will follow $2^{40}+7$ for 40 evens. The process just shown will continue indefinitely.
Here is the CJ pattern for 05 for 40 evens and the CJ pattern for $2^{40}+5=$

1099511627781 for 40 evens:

05-16-08-04-02-01-04-02-01-04-02-01-04-02-01-04-02-01-04-02-01-

04-02-01-04-02-01-04-02-01-04-02-01-04-02-01-04-02-01-04-02-01-

04-02-01-04-02-01-04-02-01-04-02-01-... .

The pattern is odd-even-even-even-even-odd-even-even-odd-even-even-odd-even-

even-odd-even-even-odd-even-even-odd-even-even-odd-even-even-odd-even-even-

odd-even-even-odd-even-even-odd-even-even-odd-even-even-odd-even-even-odd-

even-even-odd-even-even-odd-even-even-odd-even-even-odd-even-even.

Now 05+2^40=1099511627781:

1099511627781-3298534883344-1649267441672-824633720836-412316860418-

206158430209-618475290628-309237645314-154618822657-463856467972-

231928233986-115964116993-347892350980-173946175490-86973087745-

260919263236-130459631618-65229815809-195689447428-97844723714-

48922361857-146767085572-73383542786-36691771393-110075314180-

55037657090-27518828545-82556485636-41278242818-20639121409-

61917364228-30958682114-15479341057-46438023172-23219011586-
11609505793-34828517380-17414258690-8707129345-26121388036-

13060694018-6530347009-19591041028-9795520514-4897760257-

14693280772-7346640386-3673320193-11019960580-5509980290-2754990145-

8264970436-4132485218-2066242609-6198727828-3099363914.

The pattern agrees with 05 for the predicted 40 evens!

If W is any odd number, $2^{\wedge}n+W$ (n=#evens) agrees with $Z^{\wedge}+(2^{\wedge}n)+W$.
Here are a few examples:

Let's start with odd number W=3:

03-10-05-16-08-04-02-01-04-02-01-...

$5(2^{\wedge}4)+3=80+3=83(Z^{\wedge}+=5, n=4, W=3)$. This means it should agree
with 83 for 4 evens. Verifying:

83-250-125-376-188-94

It looks like they both follow the odd-even-odd-even-even-even pattern for the

first 4 evens.

Below are some examples of the symmetry of the smallest integers

with other integers using how many evens it takes 01, 03, 05, 07, 09,

11, 13, 15 to reach 1, proving any integer $x+(2^{\wedge}n)$ up to x=vm (x=01, 03, 05, 07 ,09,

11, 13 ,15, ...VM) will subcoll.

2 evens: 1-5-9-13-17-21-25-29-33-... (This is because $2^2=4$)

5 evens: 3-35-67-...($2^5=32$)

4 evens: 5-21-37-53-...($2^4=16$)

11 evens: 7-2055-...($2^{11}=2048$)

13 evens: 9-4201-...($2^{13}=4192$)

10 evens: 11-1035-...($2^{10}=1024$)

7 evens: 13-141-269-...($2^7=128$)

12 evens: 15-4111-...($2^{12}=4096$)

Choose any two odd positive integers x and y. The next two results will be $3x+1$

and $3y+1$ because of the rule for odd integers of the CJ. If y>x, the difference is

$(3y+1)-(3x+1)$. Since in our example y=31 and x=15, y-x=16. Adding x to both sides

of this equation gives us y=16+x. Substituting 16+x for y in the previous equation

gives us $(3(16+x)+1)-(3x+1)=48+3x+1-3x-1=48$. This number (48) is always 3 times

(y-x) because everything else cancels. We chose (y-x) to be a power of $2(2^n)$, making

it always even. And 3 times an even number is always even (Euclid's Elements

Book IX Proposition 28), proving agreement because even+even=even and odd+even=

odd. It will agree for n evens because dividing 3 times a power of 2 will be

even until the quotient reaches 1*3: $3(2^4)/2=3(2^3)$, $3(2^3)/2=3(2^2)$,

3(2^2)/2=3(2), 3(2)/2=3.

The continuation of the n evens from any integer to below any

VM can be shown using a hypothetical VM.

Let's choose a hypothetical VM of 1,000,000 and n evens=19 so that

$2^n=2^{19}=524,288$. We know from my proof that any integer w will

agree for 19 evens with w-(2^{19}). VM+1=1,000,000+1=

1,000,001 so let's start there:

Chosen integer	-2^19	Chosen integer-2^19
1,000,001	-524,288=	475,713
1,000,003	-524,288=	475,715
1,000,005	-524,288=	475,717
1,000,007	-524,288=	475,719
1,000,009	-524,288=	475,721
. . .		
. . .		
. . .		
1,524,281	-524,288=	999,993
1,524,283	-524,288=	999,995
1,524,285	-524,288=	999,997
1,524,287	-524,288=	999,999

All the integers in the first column agree for 19 evens with the integers in the third

column because a power of 2 (2^{19}) was subtracted from them. And half of them

subcoll in 2 evens (1,5,9,...).

Chosen integer	-2^19=	Chosen integer-2^19
1,524,289	-524,288=	1,000,001
1,524,291	-524,288=	1,000,003
1,524,293	-524,288=	1,000,005

1,524,295	-524,288=	1,000,007
. . .		
. . .		
. . .		
2,048,571	-524,288=	1,524,283
2,048,573	-524,288=	1,524,285
2,048,575	-524,288=	1,524,287
2,048,577	-524,288=	1,524,289

All the numbers above agree for 19 evens with the numbers in the list before. For

example, 1,524,293-524,288=1,000,005 can be transferred to the previous

table, using 1,000,005-524,288=475,717. And 475,717 is below the

chosen hypothetical VM 1,000,000.

Here is an idea I came up with that helps in my proof: Choose any positive whole

number below any hand-verified maximum.

Any numbers generated when applying the rules of the CJ will also reach 1
automatically.

This is because every time a number is reached it is the same thing as starting with that
number.

Let's choose 7 as an example:

7-22-11-34-17-52-26-13-40-20-10-05-16-08-04-02-01-...

Although several of the numbers reached in the process of the CJ are greater than 7,

their steps coincide with the steps of 7. The numbers greater than 7 that are reached are

22,11,34,17,52,26,13,40,20,10,16,08.

Choose any of these numbers and apply the rules of the Collatz Conjecture; it is identical to the steps of 7 if you start at that number. Choosing 52 as an example:

52-26-13-40-20-10-05-16-08-04-02-01-...

If you check above, you can see that 52 is included in the sequence for 7 and is identical

to the steps for 7 starting at 52.

Using this idea, we can start with any verified number and generate a multitude of other verified numbers. We can

double any of the numbers that are not doubled or cut-in-half already as long as we want to (ad infinitum).

We can do this because we can then use the x/2 rule to divide each of the even numbers in half (any

number doubled is even) until we reach the original array and continue as usual from there. I noticed you

can obtain another number by doubling any of the results that are the required 3x+1 after an odd number

two times. This number is also one more than a multiple of 3, proving it verified also. Every other one of

these double numbers also generates a new 3x+1 number (ad infinitum).

The reason is after the 3x+1 rule is applied and that number is doubled, we have $2(3x+1)=6x+2$ form.

For example, choose x=13, 15, and 17 as our starting positive integers. That means 3x+1 equals

40, 46, and 52 respectively. Doubling these numbers gives us 2(40)=80, 2(46)=92, 2(52)=104.

We can verify that we are 6x+2 from the starting integers by plugging in 13, 15, and 17 in

the equation 6x+2: x=13, 6(13)+2=78+2=80; x=15, 6x+2=6(15)+2=90+2=92; x=17,

6x+2=6(17)+2=102+2=104. It checks as expected. We can double it one more time, making it

2(6x+2)=12x+4. Doubling our choices generates 2(80)=160, 2(92)=184, and 2(104)=208.

The second doubling has obtained three even numbers that are one more than a multiple of 3:

160=3(53)+1, 184=3(61)+1, 208=3(69)+1.

The form 12x+4 is of the form 3x+1 because after subtracting 1 we have a number that is

divisible by 3. We have (12x+4)-1=12x+3. Noticing 12x=3(4x), it will always be divisible by 3

4x times(12x/3=4x). Now add 3 to 12x to make it the 12x+3 we derived. 12x+3 is divisible by

3 because adding 3 to any number that is divisible by 3 gives us another number that is divisible by

3. To demonstrate this we need the fact that any number y divisible by 3 will have a positive integer

x as the quotient(y/3=x or y=3x). If we add 3 to this number we have y+3=3x+3=3(x+1). The form

3(x+1) is always divisible by 3 (x+1) times. In algebra we have 3(x+1)/3=(x+1). Our form 12x+3

is divisible by three (4x+1) times. In algebra it is similar to 3x+3: 12x+3=3(4x+1), 3(4x+1)/3=4x+1.

This happens every other time we double a number in the array.

Here is an example of the effect every other doubling has on any integer:

Choose 5 as our starting integer, 3x+1=16. Let's double 16 as long as we want to:

16-32-64-128-256-512-1024-... The second doubling after 16 results are 64, 256, and 1024.

They are of the 3x+1 form because 64=3(21)+1; 256=3(85)+1; 1024=3(341)+1.

I did a computer program that proves all integers retrocoll using all the possible last three digits

of any positive integer similar to the method just explained using two digits with my brother. I explained

my method to him and he wrote the computer program. It worked perfectly, proving the Collatz

Conjecture! My brother was on the Special Effects team for the movie True Lies. It was nominated for an

Academy Award for Best Special Effects. He also did some of the Special Effects for the movie Species

and the children's television show Barney. It works because the positive integer plus one unit past any VM

will at least fall into the VM when going below itself, proving verification by the definition of VM (a

Verified Maximum is the highest number that all positive integers up to and including the VM have been

verified to by computer).

Since the patterns generated by the CJ agree according to powers of 2, every new sequence will start at

2^n+Z^+.

We have a new sequence at each 2^n+Z^+. This means we can find each new sequence by counting any

number of evens we do not already have and adding it to any Z^+, including the infinite and repeating 1-4-

2-1-4-2-1-... at the end of all integers proven to reach 1. This should help finding new sequences. I have

proved half the odd numbers will go below themselves because the first two results are even. I used the fact

that the last two digits prove the initial two results even. The last two digits of the integers are 01-05-09-

13-17-21-25-29-33-37-41-45-49-53-57-61-65-69-73-77-

81-85-89-93-97.

The idea is, for example, any integer ending in 01 will follow one of two patterns: 01-04-02 or 01-04-52.

This is so because any integer ending in 01 will have 04 as the unique even number after the 3x+1 step.

Next, the only possible two results are either 02 or 52 after the x/2 step. We have, after the first two steps,

two evens. We multiply by 3, add 1 for the 3x+1 step, then divide by 2 two times for the x/2 step. This

means we are 3x+1 after the first step, (3x+1)/2 after the second step, and (3x+1)/4 after the third step. If

we add 1 to any number greater than one after we have multiplied it by 3, it will be less than the number

obtained after we multiply it by 4. An example is 3(13)+1=40, 4(13)=52, 40<52. This means when we

divide (3x+1) by 4, we have a number less than x. All integers will agree with the integers that end in 01

for two evens if they are 01+(Z^+)2^n or 01+(Z^+)2^2=01+(Z^+)4 since n=2 for two evens.

We have proved all integers ending in 01, 05, 09, 13, 17 ...97 are approximately 3/4 of their original

starting value.

This is half of all odd numbers. Since all even numbers fall below themselves automatically due to the x/2

rule for even numbers being a division by 2, we have proved all even (1/2 the integers) and half the odd

(1/4 the integers) or 1/2+1/4=3/4 of the positive integers go below themselves after applications of the steps

of the Collatz Conjecture.

The patterns repeat as follows: Starting with 1, every power of two (2, 4, 8, 16 ,32, 64, 128, ...) determines

the distance to the next odd number with that many odds and evens in the exact order as already explained.

These patterns are fixed and will determine the pattern of any positive integer needed. The patterns will

always coincide with the ending 1-4-2-1-4-2-1-4-2-1-... pattern at least eventually. And this pattern has two

evens and one odd, bringing any chosen positive integer x approximately (3/4)x for each three steps. Let's

put it this way, start with any number past any VM of the CJ. There will be a result after each application of

the CJ. We will therefore reach the maximum number of steps it took for any of the results of the CJ below the VM. We will therefore eventually have 1-4-2-1-4-2-1-... for all sequences below any VM. We can

subtract 2^n (n=#evens) from our chosen number and have it agree for n evens.

We can choose a very high n because we are above a VM. This means our chosen integer will agree with

the 1-4-2-1-4-2-1-... for a very high number of times. This will cause a decrease of approximately 3/4 for

each three steps many times. If the very high "n" is not high enough to bring our chosen number below the

VM, we can choose more large "n"s and apply the same process indefinitely. This should bring any number

to the 1-4-2-1-4-2-1-... pattern, meaning it has reached 1.

Examples of this agreement of the 1-4-2-1-4-2-1-... pattern can be shown using any of the positive

integers as the beginning integer and any power of 2 high enough to reach the 1-4-2-1-4-2-1-... ending of

the chosen positive integer. Let's start with 3 as the starting number and 2^{16}=65,536 as the power of 2.

We claim that the odd-even patterns will agree for 16 evens including the 1-4-2-1-4-2-1-... pattern at the

end of our chosen integer 3. Let's try 3, 3+65,536=65,539, 3+2(65,536)=3+131,072=131,075,

3+3(65,536)=3+196,608=196,611. We claim that 3, 65,539, 131,075, 196,611 will agree for 16 evens, then

alternate odd-even or even-odd after the 16th even. Let's see:

3-10-05-16-08-04-02-01-04-02-01-04-02-01-04-02-01-04-02-01-04-02-01-04-02...

65539-196618-98309-294928-147464-73732-36866-18433-55300-27650-13825-41476-20738-10369-

31108-15554-7777-23332-11666-5833-17500-8750-4375-13126-6563-...

131075-393226-196613-589840-294920-147460-73730-36865-110596-55298-27649-82948-41474-

20737-62212-31106-15553-46660-23330-11665-34996-17498-8749-26248-13124-...

196611-589834-294917-884752-442376-221188-110594-55297-165892-82946-41473-

124420-

62210-31105-93316-46658-23329-69988-34994-17497-52492-26246-13123-39370-19685-...

The odd-even pattern for all four integers is odd-even-odd-even-even-even-even-odd-even-even-odd-

even-even-odd-even-even-odd-even-even-odd-even-even-odd-even-... . The pattern for the

four integers after the sixteenth even is odd-even-odd-even (02, 6563, 13124, 19685).

If G=greatest number of evens required to reach 1 up to any chosen VM and R=record delay, $2^{\wedge}G+Z^{\wedge}+$

will agree with each $Z^{\wedge}+$ for G evens. It will agree with $1+2^{\wedge}G$ for G evens, 3 will agree for $3+2^{\wedge}G$ for G

evens, 5 will agree with $5+2^{\wedge}G$ for G evens, etc. They$(2^{\wedge}G+1, 2^{\wedge}G+3, 2^{\wedge}G+5, ...2^{\wedge}G+R)$ will all go below

themselves because G has caused each of the positive integers below the VM to reach 1, a number far

below any numbers past the VM. We should choose plus one over the number of evens required for the

current delay record.

Terminology like glide, delay, completeness, etc. can be found at the site on the internet by Eric

Roosendaal, reachable through mersenne.org. He has a lot of material about the Collatz Conjecture.

I did a lot of work tinkering with the Collatz Conjecture after the first printing of this work with Peter Raeth.

We know $2^{\wedge}G+1, 5, 9, 13, 17, ...2^{\wedge}G+2n$, n=even $Z^{\wedge}+$ all fall below themselves after two evens because

they are separated by $+4=2^{\wedge}2$, proving they agree for two evens:

1-4-2-1-...

5-16-08-04-...

9-28-14-07-...

They all subcoll after the third step, separated by plus three units because of symmetry.

If we choose any positive odd integer x as our starting number, the first step will be 3x+1, an even number.

This number will be odd half the time and even half the time, alternating between the two. It can be

determined by the formula (3x+1)/2 because since 3x+1 is always even the x/2 rule is applied. If x is

verified so is (3x+1)/2.

Any number that is a power of 2 will proceed directly to 1 by successive divisions by 2(x/2 rule). Since

$2^3=8$, $2^5=32$, and $2^8=256$, we reach 1 by dividing by 2 three times for 8, five times for 32, and eight

times for 256.

Let's see:

08-04-02-01-...

32-16-08-04-02-01-...

256-128-64-32-16-08-04-02-01-...

Our prediction was correct. The sequence can be expressed algebraically by $2^n/2^p$, n=member of Z^+,

p=1, 2, 3, ..., n.

I have devised a method that gives us an algebraic method of following the results of the CJ. We start with

any chosen positive odd integer x. Applying the 3x+1 step for odd integers makes it 3x+1. Since this result

is always even, the x/2 step makes it (3x+1)/2. This result will be even half the time and odd half the time.

Here is why: Even numbers are always spaced two units apart. Let's start with 2, 4 , 6, 8, 10, ... Now divide

them by 2 to perform the x/2 rule for even integers. We have 2/2=1, 4/2=2, 6/2=3, 8/2=4, 10/2=5. We are

counting 1, 2, 3, 4, 5, proving half are odd and half are even, alternating between the two when the even

integers are spaced consecutively. This means our (3x+1)/2 integer is even half the time and odd half the time, alternating between the two. If it is odd, the algebraic equation is $3((3x+1)/2)+1$. If it is even, the

equation is $((3x+1)/2)/2=(3x+1)/4$. Two examples are $x=11$ for the first equation. We claim that since

$(3x+1)/2=(3(11)+1)/2=17$, the third result is
$3((3x+1)/2)+1=3((3(11)+1)/2)+1=3(17)+1=52$. Let's see:

11-34-17-52- are the first three results of the CJ. And 52 is the third result as claimed.

The second result for the odd numbers plus or minus two from 11 should be even because of the

predicted even-odd-even pattern. We have $x=9$, 13 as our starting integers. The second result

is $(3x+1)/2$ for $x=9$, 13. Plugging in the values, $(3(9)+1)/2=14$ and $(3(13)+1)/2=20$. The third

result will be $(3x+1)/4$ or, for $x=9$, 13, $(3(9)+1)/4=7$ and $(3(13)+1)/4=10$. Here are the sequences for

$x=9$ and $x=13$:

09-28-14-07-...

13-40-20-10-...

The results are 7 and 10 as predicted. The results will keep adding plus 3 for every plus 4 starting with

01 indefinitely. We have proved the third result is always either $(3x+1)/4$ or $3((3x+1)/2)+1$. The fourth

result has one of three possibilities: $(3x+1)/4$ will be either even or odd, algebraically this is

$(3x+1)/8$ or $3((3x+1)/4)+1$, and $3((3x+1)/2)+1$ will be even, algebraically this is $(3((3x+1)/2)+1)/2$.

An odd integer will run back into itself only if it coincides with the $x/2$ pattern for even integers

($3x+1$ always generates an even number). For example, if $x=5$, the only way for it to

repeat after the rules of the CJ are followed is if it runs into ...1280 (256x), 640 (128x), 320 (64x),

160 (32x), 80 (16x), 40 (8x), 20 (4x), 10 (2x), 5 (x). The array for 5 is:

5-16-08-04-02-01-04-02-01-...

The algebraic equations are 5(odd)-3x+1, 16 even-(3x+1)/2, 08 even-(3x+1)/4, 04 even-(3x+1)/8,

02 even-(3x+1)/16, 01 odd-(3(3x+1)/16)+1, 04 even-((3(3x+1)/16)+1)/2, 02 even-((3(3x+1)/16)+1)/4,

01 odd-(3((3(3x+1)/16)+1)/4)+1.

The patterns repeat as follows: Starting with 1, every power of two n (n=2, 4, 8, 16, 32, ...)

determines the distance to the next odd number with that many evens

and odds in the exact order (see 2^n explanation). The distance between

1 and the next pattern equivalent to oeeoe (1-4-2-1-4) is 2^3=8 because

of the 3 evens (4, 2, 4). Since 1+8=9, 9 should agree with 1 for 3 evens.

Let's see:

1-04-02-01-04

9-28-14-07-22

We can add 8 as many times as we want to--the first 3 evens will have

the same pattern. Let's add 9+8=17, try 17 now:

17-52-26-13-40

We have the same oeeoe as we had with 1 and 9. Let's add 8+17=25,

see what we get:

25-76-38-19-58

It's oeeoe as expected. It doesn't matter what number we start with,

any desired positive integer n number of evens can be found by adding

2^n (n=3, 2^n=8 in our first 4 examples) to our starting number (1,9,17,

25 in our examples). The sequences in between will follow the always

even after odd and half the time even half the time odd after an even integer (alternating odd-even or even-odd) already explained). These patterns are fixed and will determine the patterns of any positive integers needed. This pattern, determined by the patterns below, will will eventually coincide with 1-4-2-1, which is about 3-3/2-3/4 after the odd number (coinciding with 1) is multiplied by 3 with 1 added, the even number is halved (coinciding with 4), and the next number will be halved because it will be even (coinciding with 2). This means we are decreasing 3/4 for every 3 results and will eventually reach the 1-4-2-1 pattern, thereby reaching 1.

Let's put it this way--start with any number past any VM for the CJ. There will be a result after each application of the rules of the CJ. This means we will reach the maximum number of results it took for any of the integers below the VM. We therefore have all 1-4-2-1-... sequences for any number to coincide with. We can subtract 2^n (n=#evens) from our chosen number with the same pattern. We can choose a very high VM because we are above a very high VM. This means our VM will agree with the 1-4-2-1-... pattern for a very large "n" number of times. We will decrease 3/4 for every 1-4-2-... or 3 results as just shown. If the very high n is not high enough to bring our chosen number below below a VM, we can choose more large n's and apply the same process indefinitely. This should bring any number to the 1-4-2-... pattern, thereby reaching 1 because we are decreasing an infinite number of 3/4s. Quod Erat Demonstratum.

Let's choose as a hypothetical VM 100. Now our lowest

un-hypothetically verified number is 101. We should choose the

highest power of 2 that is not over 101, which is 64, or 2^6. This

means $101-64=37$ will agree with 101 for 6 evens. Let's see:

37-112-56-28-14-07-22-11-34

101-304-152-76-38-19-58-29-88

The pattern that agrees is oeeeeoeoe. The agreeing numbers will also

go below themselves proportionately--37 and 101 went below themselves

at 28, 76; 14, 38; 07, 19; 22, 38; 11, 29; and 34, 88.

The formula for the x/2 pattern that coincides with itself is $(2^n)x$, $n=Z^+$. The formula for any

step of the CJ using the number of odds y and number of evens z after any starting integer x is:

$(((3^y)x+(3^y-1))/2^z)+((3^y-2)/2^z-1)+((3^y-3)/2^z-2)+...$. The powers of 3 are y, y-1, y-2, y-3, ... and

the powers of 2 are z, z-1, z-2, ... If we set these two equations equal

to each other and prove the equation impossible, it will prove that there are not any integers

that repeat after the rules of the CJ are followed using any starting positive integer. Let's try it:

$(((3^y)x+(3^y-1))/2^z)+((3^y-2)/2^z-1)+((3^y-3)/2^z-2)+...=?(2^n)x$

$((3^y)/2^z)x+(3^y-1)/2^z+(3^y-2)/2^z-1+(3^y-3)/2^z-2+...=?(2^n)x$

Subtract $(2^n)x$ from both sides of the equation:

$((3^y)/2^z)x+(3^y-1)/2^z+(3^y-2)/2^z-1+(3^y-3)/2^z-2-(2^n)x...=?0$

Subtract all terms not containing x from both sides of the equation to get the terms containing

x on one side of the equation:

$((3^y/2^z)x-(2^n)x=?-(3^y-1)/2^z-(3^y-2)/2^z-1-(3^y-3)/2^z-2-...$

Divide through by x:

$((3^{\wedge}y/2^{\wedge}z)-(2^{\wedge}n)=?-((3^{\wedge}y-1)/2^{\wedge}z)/x-((3^{\wedge}y-2)/2^{\wedge}z-1)/x-((3^{\wedge}y-3)/2^{\wedge}z-2)/x-...$

$((3^{\wedge}y/2^{\wedge}z)-(2^{\wedge}n)=?-((3^{\wedge}y-1)/2^{\wedge}z-(3^{\wedge}y-2)/2^{\wedge}z-1-(3^{\wedge}y-3)/2^{\wedge}z-2)/x-...$Divide both sides of the equation by the

numerator of x to get x by itself:

$((3^{\wedge}y/2^{\wedge}z)-(2^{\wedge}n)/((3^{\wedge}y-1)/2^{\wedge}z-(3^{\wedge}y-2)/2^{\wedge}z-1-(3^{\wedge}y-3)/2^{\wedge}z-2)-...=?1/x$

Multiply both sides of the equation by x to get x out of the denominator:

$((3^{\wedge}y/2^{\wedge}z)-(2^{\wedge}n)/((3^{\wedge}y-1)/2^{\wedge}z-(3^{\wedge}y-2)-.../2^{\wedge}z-1-(3^{\wedge}y-3)/2^{\wedge}z-2)x=?1$

Divide both sides of the equation by the numerator and multiply both sides of the

equation by the denominator:

$x=?(3^{\wedge}y-1)/2^{\wedge}z-(3^{\wedge}y-2)/2^{\wedge}z-1-(3^{\wedge}y-3)/2^{\wedge}z-2-.../((3^{\wedge}y/2^{\wedge}z)-(2^{\wedge}n))$

This means that the fraction on the right side of the equation will have to equal an odd integer, or

$(3^{\wedge}y-1)/2^{\wedge}z-(3^{\wedge}y-2)/2^{\wedge}z-1-(3^{\wedge}y-3)/2^{\wedge}z-2-...=?D((3^{\wedge}y/2^{\wedge}z)-(2^{\wedge}n)), D=Z^{\wedge}+.$

Since there is always an even after every odd, z is at least equal to y. It doesn't appear possible

for $3^{\wedge}y/2^{\wedge}z$ to be a whole number because the numerator is always odd and the denominator is

always even. Since $2^{\wedge}n$ is always an even integer, we have a fraction minus an integer, which

cannot possibly equal an integer. Since x is any odd integer, we have proven it impossible for any

chosen positive integer to coincide with itself in any sequence of the CJ.

Let's go into some detail about the 1-4-2-1-... pattern that every positive whole number below any

VM will follow with the positive integers above this VM. The retrocoll action of the 1-4-2-1-...

pattern before a VM happens with the usual power of two agreements. We can obtain any

number of evens desired. If we need 30 evens to agree ($2^{\wedge}30$=1,073,741,824), we can choose that many

evens. And there will be several odds also.

Come to think of it, as soon as any integer that is a power of 2 number of evens from an integer that has

reached the 1-4-2-1-... pattern, it will have two evens per odd until the number of evens runs out. Since

no integer will coincide with itself, all sequences of the CJ are infinite. This means there

will always be a 3x+1 or x/2 result, ending in 1-4-2-1-... . Here are a few examples:

We can use 41 as the starting integer. The nearest power of 2 that is not

greater than 41 is $2^5=32$, it should agree with 41-32=9 for five evens:

9-28-14-07-22-11-34-17-52-26-13-40-20-10-05-16-08-04-02-01-04-02-01-04-02-01-...

41-124-62-31-94-47-142-71-214-107-322-161-484-242-121-364-182-91-274-137-412-206-

103-310-155-466-233-700-350-175-526-263-790-395-1186-593-1780-890-445-1336-668-334-

167-502-251-754-377-1132-566-283-850-425-1276-638-319-958-479-1438-719-2158-

1079-3238-1619-4858-2429-7288-3644-1822-911-2734-1367-4102-2051-6154-3077-9232-

4616-2308-1154-577-1732-866-433-1300-650-325-976-488-244-122-61-184-92-46-23-70-

35-106-53-160-80-40-20-10-05-16-08-04-02-01-...

The expected agreement to five evens pattern is odd-even-even-odd-even-odd-even-odd-even-.

The predicted disagreement after the five evens is demonstrated by the numbers 26 and

107. We are now at number 107, our new starting integer. The power of two nearest

107 without going over is $2^6=64$. We now have 107 will agree for six evens with

$107+2^6=107-64=43$:

43-130-65-196-98-49-148-74-37-112-...

The pattern of agreement is odd-even-odd-even-even-odd-even-even-odd-even-.

We have arrived at number 137. The needed 2^n is now $2^7=128$, for agreement for

seven evens with 137-128=9:

9-28-14-07-22-11-34-17-52-26-13-40-...

The pattern is odd-even-even-odd-even-odd-even-odd-even-even-odd-even-.

Our new starting whole number is 263. The highest power of two we can use is $2^8=256$. The positive whole number for agreement for eight evens is $263-256=7$. We have:

7-22-11-34-17-52-26-13-40-20-10-05-16-...

The pattern is odd-even-odd-even-odd-even-even-odd-even-even-even-odd-even-.

We have arrived at 251. Our new 2^n is $2^7=128$ for agreement for seven evens with $251-128=123$:

123-370-185-556-278-139-418-209-628-314-157-472-...

The pattern is odd-even-odd-even-even-odd-even-odd-even-even-odd-even-.

Our new number is 479. The nearest power of 2 is $2^8=256$, and it should agree for eight evens with $479-256=223$:

223-670-335-1006-503-1510-755-2266-1133-3400-1700-850-425-1276-...

The sequence's odd-even pattern is odd-even-odd-even-odd-even-odd-even-odd-even-even-even-odd-even-.

We have arrived at 1367. The lowest 2^n that is not greater than 1367 is $2^{10}=1024$. This means we expect even-odd pattern agreement for ten evens $1367-1024=343$:

343-1030-515-1546-773-2320-1160-580-290-145-436-218-109-328-164-...

The pattern is odd-even-odd-even-odd-even-even-even-even-odd-even-even-odd-even-even-.

Our next starting positive whole number is 325. The nearest power of 2 is $2^8=256$, which should agree for eight evens with $325-256=69$:

69-208-104-52-26-13-40-20-10-05-16-...

The pattern of agreement is odd-even-even-even-even-odd-even-even-even-odd-even.

We start at 35 this time. The nearest power of two is $2^5=32$. There should be the same odd-even pattern for 35 and $35-3^5=35-32=3$ for five evens:

3-10-05-16-08-04-02-...

The pattern is odd-even-odd-even-even-even-even-.

We are now at 10. Since 10 is even, we perform the x/2 rule until we reach an odd number. It looks like 10/2=05, let's start there. The nearest power of 2 without going over is 2^2=4. We should have agreement with 5 and 5-2^2=5-4=1 for two evens.

01-04-02-...

The pattern is odd-even-even-.

We have reached an even number (4). We use the x/2 rule twice to reach 1:

4-2-1-.

We have reached 1!

With these new ideas I can predict what happens to any odd positive integer, including any above the VM, no matter how much higher. Let's start with the record delay(R) (the number of steps it takes for it to reach 1), which is 17,026,512,240,355,369. It took 2,042 steps for it to reach 1. Any power of 2(2^n) plus R will agree for n evens. This means any number that is a multiple of (2^(2,042)+R) will subcoll.

All integers x starting at (2^(2,042)+x) will follow the 1-4-2-1 sequence, bringing it down (3/4)x for every two evens past (2^(2,042)+x), if not before.

Starting just over the delay record R, the same power of two idea applies. As an illustration, since 2^14=16,384, any multiple of (16,384+R) will agree for 14 evens with R.

The Collatz Conjecture has been verified up to 1,200,000,000,000(one trillion two hundred billion).

The first integer that has not been verified is the odd number 1,200,000,000,001 (one trillion two hundred billion and one). All the properties claimed still apply. We can choose any power of two(2^n) and expect agreement for n evens. Let's choose 2 evens as a

starter. Every positive whole number +4 or a multiple of +4(4,8,12,16,20,24,28,32,...) will

agree with 1,200,000,000,001 for two evens. Here are a few samples:

1,200,000,000,001-3,600,000,000,004-1,800,000,000,002-900,000,000,001

This sample and the ones that follow have the two-digit endings predicted in the tables as do any chosen positive integers.

Since the pattern is odd-even-even-, the third step will always fall below the original starting integer. The number after the third step in our first example is 900,000,000,001, which is below the starting number (1,200,000,000,001). Continuing:

1,200,000,000,005-3,600,000,000,016-1,800,000,000,008-900,000,000,004-

We added plus four to our original starting number, the odd-even-even- sequence has prevailed! We also have the odd-even disagreement after the third step. The integers that are opposites are 900,000,000,001 and 900,000,000,004. The third result is also less than the starting integer of the sequence as expected. The third step will be +3 units per 4 units that is added to the starting number. This illustrates my explanation that one odd and two evens is always the equation (3x+1)/4. Plugging in either of our starting integers will get us the third step. Let's try 1,200,000,000,001 for x. In this case 3x is 3,600,000,000,003. And 3x+1 is 3,600,000,000,004. Dividing by 4 gives

us 900,000,000,001, the third result! Here are a few more:

1,200,000,000,009-3,600,000,000,028-1,800,000,000,014-900,000,000,007-
1,200,000,000,013-3,600,000,000,040-1,800,000,000,020-900,000,000,010-
1,200,000,000,017-3,600,000,000,052-1,800,000,000,026-900,000,000,013-
1,200,000,000,021-3,600,000,000,064-1,800,000,000,032-900,000,000,016-
1,200,000,000,025-3,600,000,000,076-1,800,000,000,038-900,000,000,019-
1,200,000,000,029-3,600,000,000,088-1,800,000,000,044-900,000,000,022-
1,200,000,000,033-3,600,000,000,100-1,800,000,000,056-900,000,000,025-

We can easily see by observation several important things. The first result is 3x+1 because we chose odd integers to start with. This causes a uniform twelve unit increase because, since we are increasing +4 for each starting positive integer x, 3x+1 increases +12. We can demonstrate this in algebra as follows:

Let the original positive whole number be x, the next integer y will be y=x+4 because we have chosen them four units apart. The equation for y is 3y+1 because y is odd. Substituting

y=x+4 for y gives us y=3(x+4)+1=3x+13. Subtract 3x+1 from 3x+13 to find the distance between

them. We have 3x+13-3x+1=12, the expected distance. Since 3x+1 is always even, the second

result is (3x+1)/2. The starting integer is still x and the next integer is still y=x+4. The new

equation for y is (3y+1)/2. Substituting y=x+4 in the equation for y gets (3(x+4)+1)/2=(3x+12+1)/2=

(3x+13)/2. Subtracting (3x+1)/2 from (3x+13)/2 as before: (3x+13)/2-(3x+1)/2=(3x+13-3x-1)/2=

12/2=6, the expected number of units. The third results after the odd-even-even pattern are all

below the original starting positive number as predicted. We have proved that 1,200,000,000,001 and

every integer any multiple of +4 will decrease to below itself after the symmetrical odd-even-even-

pattern ad infinitum.

The next integer in the straticoll is 1,200,000,000,003, we can use the same odd-even pattern

agreement as usual.

1,200,000,000,003-3,600,000,000,010-1,800,000,000,005-5,400,000,000,016-

2,700,000,000,008-1,350,000,000,004-675,000,000,002. The number has

subcolled already. It took five evens, so the same odd-even pattern will

occur every 2^5=+32 units. It will be the same odd-even pattern as 03

(odd-even-odd-even-even-even-even-).

Let's add a high number that is a muliple of 32, like 32,000,000=1,000,000x32.

Our new starting number is 32,000,000+1,200,000,000,003=1,200,032,000,003.

It should have the same odd-even pattern for five evens:

1,200,032,000,003-3,600,096,000,010-1,800,048,000,005-5,400,144,000,016-

2,700,072,000,008-1,350,036,000,004-675,018,000,002-.

We have the identical odd-even-odd-even-even-even-even- pattern.

Now let's try multiplying a high number without too many zeros times 32, add it to 1,200,000,000,003

and expect agreement for 5 evens (2^5=32):

32x12349098476873=395,171,151,259,936+1,200,000,000,003=396,371,151,259,939.

396,371,151,259,939-1,189,113,453,779,818-594,556,726,889,909-
1,783,670,180,669,728-

891,835,090,334,864-445,917,545,167,432-222958772583716-.

It can be shown algebraically that all integers reach 1 after a finite number of applications of

the rules of the Collatz Conjecture are applied, which follows.

If we can prove all odd integers reach 1, all even integers will reach either 1 or an odd number,

thereby proving the CJ. The odd starting integer will require the 3x+1 rule for the CJ. This

means x can equal 1, 3, 5, ...(any Z^+). If 3x+1=2^n, it will subcoll to 1(...32-16-08-04-02-01...).

If x=1, 3x+1=4=2^2. We well know that the CJ sequence for 1 is 1-4-2-1. Next, if x=5,

3x+1=16=2^4. Once again, the CJ sequence is 5-16-08-04-02-01. If x=1, 3, 5, 7,...,
3x+1=

4, 10, 16, 22, ..., increasing by +6. The equation for this sequence is 6n+4, n=0, 1 ,2, 3,.... .

We have a power of 2 at 4x1=4, 4x4=16, 4x4x4=64, 4x4x4x4=128, ... Since the result after

3x+1 is always even, the equation is (3x+1)/2. The result after (3x+1)/2 is even half the time

and odd half the time, alternating between the two. If it is odd, the new equation is 3((3x+1)/2)+1.

The equation (3x+1)/2 is odd when x=3, 7 , 11, 15, ... When x=1, 5, 9, 13,... (3x+1)/2 is even,

so the x/2 rule for even integers applies.

It can be shown algebraically that all integers reach 1 after a finite number of applications of the rules of the Collatz Conjecture are applied, which follows.

If we can prove all odd integers reach 1, all even integers will reach either 1 or an odd number, thereby proving the CJ. The odd starting integer will require the 3x+1 rule for the CJ. This means x can equal 1, 3, 5, ...(any $Z^\wedge+$). If $3x+1=2^\wedge n$, it will subcoll to 1(...32-16-08-04-02-01...).

If x=1, $3x+1=4=2^\wedge 2$. We well know that the CJ sequence for 1 is 1-4-2-1. Next, if x=5, $3x+1=16=2^\wedge 4$. Once again, the CJ sequence is 5-16-08-04-02-01. If x=1, 3, 5 ,7, ..., $3x+1=4$, 10, 16 , 22, ..., increasing by +6. The equation for this sequence is 6n+4, n=0, 1, 2, 3, ...

We have a power of 2 at 4x1=4, 4x4=16, 4x4x4=64, 4x4x4x4=128, ... Since the result after 3x+1 is always even, the equation is (3x+1)/2. The result after (3x+1)/2 is even half the time and odd half the time, alternating between the two. If it is odd, the new equation is 3((3x+1)/2)+1.

The equation (3x+1)/2 is odd when x=3, 7, 11, 15, ... When x=1, 5, 9, 13, ... (3x+1)/2 is even, so the x/2 rule for even integers applies. For x=3, 3((3x+1)/2)+1=16=2^\wedge 4. The results after x=3, 7, 11, 15 are 16, 34, 52, 70 respectively, increasing by +18.

Let's use the number of evens and number of odds and make the equation equal to 1. Starting with x=1, we have two evens and one odd before we reach 1(1-4-2-1). The equations in order are 1 odd, (3x+1)=4 even, (3x+1)/2=2 even, (3x+1)/4=1. Our equation for the starting integer 1 is (3x+1)/4.

Since $2^\wedge n$ in this case is $2^\wedge 2$ because of the two evens, all integers +2^2=+4 from 1 will agree for two evens, proving they subcoll. Testing 1, 5, 9, 13 with the equation (3x+1)/4, we get 1 , 4, 7, 10 respectively, all subcoll at these integers and with the same odd-even-even pattern. Because of the symmetric odd-even and even-odd or even-even patterns, the +3 interval will continue indefinitely, and will always subcoll as predicted. Next, let's try x= 03 as the starting positive integer. We have 03-10-05-16-08-04-02-01, or two odds and five evens. The equations are 3 odd, (3x+1), 10 even (3x+1)/2, 5 odd, 3((3x+1)/2)+1,

16 even, 3 ((3x+1)/2)+1)/2, 8 even,

3((3x+1)/2)+1)/4, 4 even, 3((3x+1)/2)+1)/8, 2 even, 3((3x+1)/2)+1)/16=1. Analogous to

1, since 3 has five evens, all integers 3+n(2^5)(n=0,1,2,3,4,...)=3+32n=3, 35, 67, 99...

will subcoll at that point. Checking by plugging in x=35 ,67, 99 in the equation

3((3x+1)/2)+1)/16=10, 38, 56, they all subcoll exactly at the predicted integers and have

the same odd-even-odd-even-even-even-even pattern. Evidently this proves any starting

integer will subcoll because of the odd-even and even-odd or even-even alternating

patterns and 2^n agreement. And if every integer subcolls, it means

VM+1 will fall into the VM, thereby proving it reaches 1. The same reasoning goes for

VM+2, 3, 4, ... if applied at each +1 interval.

We can demonstrate that the number of integers that could possibly not reach 1 can be

made as small a number as possible using the 2^n agreement. The first integer to use is

01. It has one odd and two evens before it reaches 1.

This means the 2^n agreement is 2^2=4, so every integer +4 from 1 (1, 5, 9,...) agrees

with 1 until it reaches 1, proving they subcoll. Also, we are only considering the odd

integers, so 1/4/2=1/2 of the odd integers subcoll due to following

the pattern of 01. Next, 03 takes 5 evens, or 2^n=2^5=32. Analogous to 01, 03 makes

1/32/2=1/16 odd integers subcoll.

Now 05 makes 1/16/2=1/8 odd integers subcoll because it takes 4 evens for it to reach 1.

Adding the fractions, we have already proven 1/2+1/16+1/8=0.5+0.125+0.0625=0.6875

will subcoll. We can continue this process to as close to 1.000... as we want to. This is

because the number of 2^n agreement not accounted for will keep decreasing due to the

fact that, for example, if 2^n=2^5=32, any integer x over 32 will have to agree with x-32,

eventually making most of them agree.

Let's try a very large integer to show what I mean. The positive integer

370000078956700004567 is well over the current VM, its CJ pattern for the first five
evens:

370000078956700004567-1110000236870100013702-555000118435050006851-

1665000355305150020554-832500177652575010277-2497500532957725030832-

1248750266478862515416-624375133239431257708.

The pattern is odd-even-odd-even-odd-even-even-even.

Adding +32 to 370000078956700004567 gives us 370000078956700004599. The CJ

pattern for the first five evens:

370000078956700004599-1110000236870100013798-555000118435050006899-

1665000355305150020698-832500177652575010359-2497500532957725031048-

1248750266478862515524-624375133239431257762.

The pattern agrees for five evens as predicted: odd-even-odd-even-odd-even-even-even.

In a nutshell, we know that the 2^n agreement between the starting integers causes
agreement for n evens. All starting integers

above a VM of two million or more will agree with for 20 evens because 2^20=1048576.
If we start at the point where

all starting integers below the VM have reached 1, all these integers have the 1-4-2-1-4-
2-1-4-2-1-... pattern. This means

all starting integers above the VM, after the number of steps it has taken the longest
sequence below the VM to reach 1, will

follow the 1-4-2-1-4-2-1-4-2-1-... pattern for as many multiples of 20 steps as needed,
thereby dereasing 3/4 per odd-even-even and reaching 1. If it is

more than one million above the VM, Z^+*(VM) will have the same pattern agreement.

The Collatz Conjecture is now a theorem!!!

Chapter 2

The Collatz sequences for the integers 01 to 231:

01-04-02-01-04-02-01-04-02-01-04-02-01-04-02-01-04-02-01-04-02-01-...

03-10-05-16-08-04-02-01-04-02-01-04-02-01-04-02-01-04-02-01-...

05-16-08-04-02-01-04-02-01-04-02-01-04-02-01-04-02-01-...

07-22-11-34-17-52-26-13-40-20-10-05-16-08-04-02-01-04-02-01-...

09-28-14-07-22-11-34-17-52-26-13-40-20-10-05-16-08-04-02-01-04-02-01-...

11-34-17-52-26-13-40-20-10-05-16-08-04-02-01-04-02-01-04-02-01-...

13-40-20-10-05-16-08-04-02-01-04-02-01-04-02-01-04-02-01-04-02-01-...

15-46-23-70-35-106-53-160-80-40-20-10-05-16-08-04-02-01-04-02-01-...

17-52-26-13-40-20-10-05-16-08-04-02-01-04-02-01-04-02-01-...

19-58-29-88-44-22-11-34-17-52-26-13-40-20-10-05-16-08-04-02-01-04-02-01-...

21-64-32-16-08-04-02-01-04-02-01-04-02-01-04-02-01-04-02-01-...

23-70-35-106-53-160-80-40-20-10-05-16-08-04-02-01-04-02-01-04-02-01-...

25-76-38-19-58-29-88-44-22-11-34-17-52-26-13-40-20-10-05-16-08-04-02-01-
04-02-01-04-02-01-04-02-01-04-02-01-...

27-82-41-124-62-31-94-47-142-71-214-107-322-161-484-242-121-364-182-91-274-
137-412-206-103-310-155-466-233-700-350-175-526-263-790-395-1186-593-1780-
890-445-1336-668-334-167-502-251-754-377-1132-566-283-850-425-1276-638-319-
958-479-1438-719-2158-1079-3238-1619-4858-2429-7288-3644-1822-911-2734-1367-
4102-2051-6154-3077-9232-4616-2308-1154-577-1732-866-433-1300-650-325-976-
488-244-122-61-184-92-46-23-70-35-106-53-160-80-40-20-10-05-16-08-04-02-01-
04-02-01-04-02-01-04-02-01-...

29-88-44-22-11-34-17-52-26-13-40-20-10-05-16-08-04-02-01-04-02-01-04-02-01-...

31-94-47-142-71-214-107-322-161-484-242-121-364-182-91-274-137-412-206-
103-310-155-466-233-700-350-175-526-263-790-395-1186-593-1780-890-445-1336-
668-334-167-502-251-754-377-1132-566-283-850-425-1276-638-319-
958-479-1438-719-2158-1079-3238-1619-4858-2429-7288-3644-1822-911-2734-1367-
4102-2051-6154-3077-9232-4616-2308-1154-577-1732-866-433-1300-
650-325-976-488-244-122-61-184-92-46-23-70-35-106-53-160-80-40-20-10
-05-16-08-04-02-01-04-02-01-04-02-01-04-02-01-04-02-01-...

33-100-50-25-76-38-19-58-29-88-44-22-11-34-17-52-26-13-40-20-10-05-16-
08-04-02-01-04-02-01-04-02-01-04-02-01-04-02-01-...

35-106-53-160-80-40-20-10-05-16-08-04-02-01-04-02-01-04-02-01-04-02-01-
04-02-01-...

37-112-56-28-14-07-22-11-34-17-52-26-13-40-20-10-05-16-08-04-02-01-
04-02-01-04-02-01-...

39-118-59-178-89-268-134-67-202-101-304-152-76-38-19-58-29-88-44-22-11-
34-17-52-26-13-40-20-10-05-16-08-04-02-01-04-02-01-04-02-01-...

41-124-62-31-94-47-142-71-214-107-322-161-484-242-121-364-182-91-274-
137-412-206-103-310-155-466-233-700-350-175-526-263-790-395-1186-593-
1780-890-445-1336-668-334-167-502-251-754-377-1132-566-283-850-425-
1276-638-319-958-479-1438-719-2158-1079-3238-1619-4858-2429-7288-3644-
1822-911-2734-1367-4102-2051-6154-3077-9232-4616-2308-1154-577-1732-
866-433-1300-650-325-976-488-244-122-61-184-92-46-23-70-35-106-53-160-80-40-20-
10-05-16-08-04-02-01-04-02-01-04-02-01-...

43-130-65-196-98-49-148-74-37-112-56-28-14-07-22-11-34-17-52-26-13-40-
20-10-05-16-08-04-02-01-04-02-01-04-02-01-...

45-136-68-34-17-52-26-13-40-20-10-05-16-08-04-02-01-04-02-01-04-02-01-...

47-142-71-214-107-322-161-484-242-121-364-182-91-274-137-412-206-103-
310-155-466-233-700-350-175-526-263-790-395-1186-593-1780-890-445-1336-
668-334-167-502-251-754-377-1132-566-283-850-425-1276-638-319-958-479-
1438-719-2158-1079-3238-1619-4858-2429-7288-3644-1822-911-2734-1367-
4102-2051-6154-3077-9232-4616-2308-1154-577-1732-866-433-1300-650-325-
976-488-244-122-61-184-92-46-23-70-35-106-53-160-80-40-20-10-05-16-08-04-02-01-
04-02-01-04-02-01-04-02-01-...

49-148-74-37-112-56-28-14-07-22-11-34-17-52-26-13-40-20-10-05-16-08-04-02-01-04-
02-01-04-02-01-...

51-154-77-232-116-58-29-88-44-22-11-34-17-52-26-13-40-20-10-05-16-08-04-02-01-
04-02-01-04-02-01-...

53-160-80-40-20-10-05-16-08-04-02-01-04-02-01-04-02-01-...

55-166-83-250-125-376-188-94-47-142-71-214-107-322-161-484-242-121-364-
182-91-274-137-412-206-103-310-155-466-233-700-350-175-526-263-790-395-
1186-593-1780-890-445-1335-668-334-167-502-251-754-377-1132-566-283-
850-425-1276-638-319-958-479-1438-719-2158-1079-3238-1619-4858-2429-
7288-3644-1822-911-2734-1367-4102-2051-6154-3077-9232-4616-2308-1154-
577-1732-866-433-1300-650-325-976-488-244-122-61-184-92-46-23-70-35-106-

53-160-80-40-20-10-05-16-08-04-02-01-04-02-01-04-02-01-...

57-172-86-43-130-65-196-98-49-148-74-37-112-56-28-14-07-22-11-34-17-52-26-13-40-20-10-05-16-08-04-02-01-04-02-01-04-02-01-...

59-178-89-268-134-67-202-101-304-152-76-38-19-58-29-44-22-11-34-17-52-26-13-40-20-10-05-16-08-04-02-01-04-02-01-04-02-01-...

61-184-92-46-23-70-35-106-53-160-80-40-20-10-05-16-08-04-02-01-04-02-01-04-02-01-...

63-190-95-286-143-430-215-646-323-970-485-1456-728-364-182-91-274-147-442-221-664-332-166-83-250-125-376-188-94-47-142-71-214-107-322-161-484-242-121-364-182-91-274-137-412-206-103-310-155-466-233-700-350-175-526-263-790-395-1186-593-1780-890-445-1336-668-334-167-502-251-754-377-1132-566-283-850-425-1276-638-319-958-479-1438-719-2158-1079-3238-1619-4858-2429-7288-3644-1822-911-2734-1367-4102-2051-6154-3077-9232-4616-2308-1154-577-1732-866-433-1300-650-325-976-488-244-122-61-184-92-46-23-70-35-106-53-160-80-40-20-10-05-16-08-04-02-01-04-02-01-04-02-01-...

65-196-98-49-148-74-37-112-56-28-14-07-22-11-34-17-52-26-13-40-20-10-05-16-08-04-02-01-04-02-01-04-02-01-...

67-202-101-304-152-76-38-19-58-29-88-44-22-11-34-17-52-26-13-40-20-10-05-16-08-04-02-01-04-02-01-...

69-208-104-52-26-13-40-20-10-05-16-08-04-02-01-04-02-01-04-02-01-...

71-214-107-322-161-484-242-121-364-182-91-274-137-412-206-103-310-155-466-233-700-350-175-526-263-790-395-1186-593-1780-890-445-1336-668-334-167-502-251-754-377-1132-566-283-850-425-1276-638-319-958-479-1438-719-2158-1079-3238-1619-4858-2429-7288-3644-1822-911-2734-1367-4102-2051-6154-3077-9232-4616-2308-1154-577-1732-866-433-1300-650-325-976-488-244-122-61-184-92-46-23-70-35-106-53-160-80-40-20-10-05-16-08-04-02-01-04-02-01-04-02-01-...

73-220-110-55-166-83-250-125-376-188-94-47-142-71-214-107-322-161-484-242-121-364-182-91-274-137-412-206-103-310-155-466-233-700-350-175-526-263-790-395-1186-593-1780-890-445-1336-668-334-167-502-151-754-377-1132-566-283-850-425-1276-638-319-958-479-1438-719-2158-1079-3238-1619-4858-2429-7288-3644-1822-911-2734-1367-4102-2051-6154-3077-9232-4616-2308-1154-577-1732-866-433-1300-650-325-976-488-244-122-61-184-92-46-23-70-35-106-53-160-80-40-20-10-05-16-08-04-02-01-04-02-01-...

75-226-113-340-170-85-256-128-64-32-16-08-04-02-01-04-02-01-04-02-01-...

77-232-116-58-29-88-44-22-11-34-17-52-26-13-40-20-10-05-16-08-04-02-01-
04-02-01-04-02-01-...

79-238-119-358-179-538-269-808-404-202-101-304-152-76-38-19-58-29-88-
44-22-11-34-17-52-26-13-40-20-10-05-16-08-04-02-01-04-02-01-04-02-01-...

81-244-122-61-184-92-46-23-70-35-106-53-160-80-40-20-10-05-16-08-04-02-01-04-02-
01-04-02-01-...

83-250-125-376-188-94-47-142-71-214-107-322-161-484-242-242-121-364-182-91-274-
137-412-206-103-310-155-466-233-700-350-175-526-263-790-395-1186-
593-1780-890-445-1336-668-334-167-502-251-754-377-1132-566-283-850-425-
1276-638-319-958-479-1438-719-2158-1079-3238-1619-4858-2429-7288-3644-
1822-911-2734-1367-4102-2051-6154-3077-9232-4616-2308-1154-577-1732-
866-433-1300-650-325-976-488-244-122-61-184-92-46-23-70-35-106-53-160-80-40-20-
10-05-16-08-04-02-01-04-02-01-04-02-01-04-02-01-...

85-256-128-64-32-16-08-04-02-01-04-02-01-04-02-01-...

87-262-131-394-197-592-296-148-74-37-112-56-28-14-07-22-11-34-17-52-26-13-40-20-
10-05-16-08-04-02-01-04-02-01-04-02-01-...

89-268-134-67-67-202-101-304-152-76-38-19-58-29-88-44-22-11-34-17-52-
26-13-40-20-10-05-16-08-04-02-01-04-02-01-04-02-01-...

91-274-137-412-206-103-310-155-466-233-700-350-175-526-263-790-395-
1186-593-1780-890-445-1336-668-334-167-502-251-754-377-1132-566-283-
850-425-1276-638-319-958-479-1438-719-2158-1079-3238-1619-4858-2429-
7288-3644-1822-911-2734-1367-4102-2051-6154-3077-9232-4616-2308-1154-
577-1732-866-433-1300-650-325-976-488-244-122-61-184-92-46-23-70-35-106-
53-160-80-40-20-10-05-16-08-04-02-01-04-02-01-04-02-01-...

93-280-140-70-35-106-53-160-80-40-20-10-05-16-08-04-02-01-04-02-01-...

95-286-143-430-215-646-323-970-485-1456-728-364-182-91-274-137-412-206-
103-310-155-466-233-700-350-175-526-263-790-395-1186-593-1780-890-445-
1336-668-334-167-502-251-754-377-1132-566-283-850-425-1276-638-319-958-
479-1438-719-2158-1079-3238-1619-4858-2429-7288-3644-1822-911-2734-
1367-4102-2051-6154-3077-9232-4616-2308-1154-577-1732-866-433-1300-650-
325-976-488-244-122-61-184-92-46-23-70-35-106-53-160-80-40-20-10-05-16-08-04-02-

01-04-02-01-04-02-01-...

97-292-146-73-220-110-55-166-83-250-125-376-188-94-47-142-71-214-107-322-161-
484-242-121-364-182-91-274-137-412-206-103-310-155-466-233-
700-350-175-526-263-790-395-1186-593-1780-890-445-1336-668-334-167-502-
251-754-377-1132-566-283-850-425-1276-638-319-958-479-1438-719-2158-1079-3238-
1619-4858-2429-7288-3644-1822-911-2734-1367-4102-2051-6154-3077-9232-4616-
2308-1154-577-1732-866-433-1300-650-325-976-488-244-122-
61-184-92-46-23-70-35-106-53-160-80-40-20-10-05-16-08-04-02-01-04-02-01-
04-02-01-...

99-298-149-448-224-112-56-28-14-07-22-11-34-17-52-26-13-40-20-10-05-16-
08-04-02-01-04-02-01-04-02-01-...

101-304-152-76-38-19-58-29-88-44-22-11-34-17-52-26-13-40-20-10-05-16-08-04-02-
01-04-02-01-04-02-01-04-02-01-...

103-310-155-466-233-700-350-175-526-263-790-395-1186-593-1780-890-445-
1336-668-334-167-502-251-754-377-1132-566-283-850-425-1276-638-319-
958-479-1438-719-2158-1079-3238-1619-4858-2429-7288-3644-1822-911-
2734-1367-4102-2051-6154-3077-9232-4616-2308-1154-577-1732-866-433-
1300-650-325-976-488-244-122-61-184-92-46-23-70-35-106-53-160-80-40-20-10-05-
16-08-04-02-01-04-02-01-...

105-316-158-79-238-119-358-179-538-269-808-404-202-101-304-152-76-38-
19-58-29-88-44-22-11-34-17-52-26-13-40-20-10-05-16-08-04-02-01-04-02-01-04-02-01-
...

107-322-161-484-242-121-364-182-91-274-137-412-206-103-310-155-466-233-
700-350-175-526-263-790-395-1186-593-1780-890-445-1336-668-334-167-502-
251-754-377-1132-566-283-850-425-1276-638-319-958-479-1438-719-2158-
1079-3238-1619-4858-2429-7288-3644-1822-911-2734-1367-4102-2051-6154-
3077-9232-4616-2308-1154-577-1732-866-433-1300-650-325-976-488-244-122-
61-184-92-46-23-70-35-106-53-160-80-40-20-10-05-16-08-04-02-01-04-02-01-
04-02-01-...

109-328-164-82-41-124-62-31-94-47-142-71-214-107-322-161-484-242-121-
364-182-91-274-137-412-206-103-310-155-466-233-700-350-175-526-263-
790-395-1186-593-1780-890-445-1336-668-334-167-502-251-754-377-1132-
566-283-850-425-1276-638-319-958-479-1438-719-2158-1079-3238-1619-4858-
2429-7288-3644-1822-911-2734-1367-4102-2051-6154-3077-9232-4616-2308-1154-

577-1732-866-433-1300-650-325-976-488-244-122-61-184-92-46-23-70-
35-106-53-160-80-40-20-10-05-16-08-04-02-01-04-02-01-04-02-01-...

111-334-167-502-251-754-377-1132-566-283-850-425-1276-638-319-958-
479-1438-719-2158-1079-3238-1619-4858-2429-7288-3644-1822-911-2734-
1367-4102-2051-6154-3077-9232-4616-2308-1154-577-1732-866-433-1300-
650-325-976-488-244-122-61-184-92-46-23-70-35-106-53-160-80-40-20-10-05-16-08-
04-02-01-04-02-01-...

113-340-170-85-256-128-64-32-16-08-04-02-01-04-02-01-04-02-01-...

115-346-173-520-260-130-65-196-98-49-148-74-37-112-56-28-14-07-22-11-34-
17-52-26-13-40-20-10-05-16-08-04-02-01-04-02-01-04-02-01-...

117-352-176-88-44-22-11-34-17-52-26-13-40-20-10-05-16-08-04-02-01-
04-02-01-04-02-01-...

119-358-179-538-269-808-404-202-101-304-152-76-38-19-58-29-88-44-22-11-
34-17-52-26-13-40-20-10-05-16-08-04-02-01-04-02-01-04-02-01-...

121-364-182-91-274-137-412-206-103-310-155-466-233-700-350-175-526-263-790-
395-1186-593-1780-890-445-1336-668-334-167-502-251-754-377-1132-566-283-850-
425-1276-638-319-958-479-1438-719-2158-1079-3238-1619-4858-2429-
7288-3644-1822-911-2734-1367-4102-2051-6154-3077-9232-4616-2308-1154-
577-1732-866-433-1300-650-325-976-488-244-122-61-184-92-46-23-70-35-106-
53-160-80-40-20-10-05-16-08-04-02-01-04-02-01-04-02-01-...

123-370-185-556-278-139-418-209-628-314-157-472-236-118-59-178-89-268-
134-67-202-101-304-152-76-38-19-58-29-88-44-22-11-34-17-52-26-13-40-20-
10-05-16-08-04-02-01-04-02-01-04-02-01-...

125-376-188-94-47-142-71-214-107-322-161-484-242-121-364-182-91-274-137-
412-206-103-310-155-466-233-700-350-175-526-263-790-395-1186-593-
1780-890-445-1336-668-334-167-502-251-754-377-1132-566-283-850-425-1276-638-
319-958-479-1438-719-2158-1079-3238-1619-4858-2429-7288-3644-1822-
911-2734-1367-4102-2051-6154-3077-9232-4616-2308-1154-577-1732-866-
433-1300-650-325-976-488-244-122-61-184-92-46-23-70-35-106-53-160-80-40-20-10-
05-16-08-04-02-01-04-02-01-04-02-01-...

127-382-191-574-287-862-431-1294-647-1942-971-2914-1457-4372-2186-
1093-3280-1640-820-410-205-616-308-154-77-232-116-58-29-88-44-22-11-
34-17-52-26-13-40-20-10-05-16-08-04-02-01-04-02-01-04-02-01-...

129-388-194-97-292-146-73-220-110-55-166-83-250-125-376-188-94-47-
142-71-214-107-322-161-484-242-121-364-182-91-274-137-412-206-103-

310-155-466-233-700-350-175-526-263-790-395-1186-593-1780-890-445-1336-668-
334-167-502-251-754-377-1132-566-283-850-425-1276-638-319-958-479-
1438-719-2158-1079-3238-1619-4858-2429-7288-3644-1822-911-2734-1367-4102-
2051-6154-3077-9232-4616-2308-1154-577-1732-866-433-1300-650-
325-976-488-244-122-61-184-92-46-23-70-35-106-53-160-80-40-20-10-05-
16-08-04-02-01-04-02-01-04-02-01-...

131-394-197-592-296-148-74-37-112-56-28-14-07-22-11-34-17-52-26-13-40-20-10-05-
16-08-04-02-01-04-02-01-04-02-01-...

133-400-200-100-50-25-76-38-19-58-29-88-44-22-11-34-17-52-26-13-40-20-10-05-16-
08-04-02-01-04-02-01-04-02-01-...

135-406-203-610-305-916-458-229-688-344-172-86-43-130-65-196-98-49-
148-74-37-112-56-28-14-07-22-11-34-17-52-26-13-40-20-10-05-16-08-04-
02-01-04-02-01-04-02-01-...

137-412-206-103-310-155-466-233-700-350-175-526-263-790-395-1186-593-1780-890-
445-1336-668-334-167-502-251-754-377-1132-566-283-850-425-
1276-638-319-958-479-1438-719-2158-1079-3238-1619-4858-2429-7288-
3644-1822-911-2734-1367-4102-2051-6154-3077-9232-4616-2308-1154-
577-1732-866-433-1300-650-325-976-488-244-122-61-184-92-46-23-70-35-
106-53-160-80-40-20-10-05-16-08-04-02-01-04-02-01-04-02-01-...

139-418-209-628-314-157-472-236-118-59-178-89-268-134-67-202-101-
304-152-76-38-19-58-29-88-44-22-11-34-17-52-26-13-40-20-10-05-16-08-04-02-01-04-
02-01-04-02-01

141-424-212-106-53-160-80-40-20-10-05-16-08-04-02-01-04-02-01-04-02-01-
04-02-01-...

143-430-215-646-323-970-485-1456-728-364-182-91-274-137-412-206-103-310-
155-466-233-700-350-175-526-263-790-395-1186-593-1780-890-445-1336-
668-334-167-502-251-754-377-1132-566-283-850-425-1276-638-319-958-479-
1438-719-2158-1079-3238-1619-4858-2429-7288-3644-1822-911-2734-1367-4102-
2051-6154-3077-9232-4616-2308-1154-577-1732-866-433-1300-650-325-976-488-244-
122-61-184-92-46-23-70-35-106-53-160-80-40-20-10-05-16-08-04-02-01-04-02-01-04-
02-01-...

145-436-218-109-328-164-82-41-124-62-31-94-47-142-71-214-107-322-161-484-242-
121-364-182-91-274-137-412-206-103-310-155-466-233-700-350-175-
526-263-790-395-1186-593-1780-890-445-1336-668-334-167-502-251-754-377-
1132-566-283-850-425-1276-638-319-958-479-1438-719-2158-1079-3238-1619-

4858-2429-7288-3644-1822-911-2734-1367-4102-2051-6154-3077-9232-4616-
2308-1154-577-1732-866-433-1300-650-325-976-488-244-122-61-184-92-46-23-
70-35-106-53-160-80-40-20-10-05-16-08-04-02-01-04-02-01-04-02-01-...

147-442-221-664-332-166-83-250-125-376-188-94-47-142-71-214-107-322-
161-484-242-121-364-182-91-274-137-412-206-103-310-155-466-233-700-350-
175-526-263-790-395-1186-593-1780-890-445-1336-668-334-167-502-251-754-
377-1132-566-283-850-425-1276-638-319-958-479-1438-719-2158-1079-3238-
1619-4858-2429-7288-3644-1822-911-2734-1367-4102-2051-6154-3077-9232-
4616-2308-1154-577-1732-866-433-1300-650-325-976-488-244-122-61-184-
92-46-23-70-35-106-53-160-80-40-20-10-05-16-08-04-02-01-04-02-01-04-02-01-...

149-448-224-112-56-28-14-07-22-11-34-17-52-26-13-40-20-10-05-16-08-04-02-01-04-
02-01-04-02-01-...

151-454-227-682-341-1024-512-256-128-64-32-16-08-04-02-01-04-02-01-04-02-01-04-
02-01-...

153-460-230-115-346-173-520-260-130-65-196-98-49-49-148-74-37-112-56-28-14-07-
22-11-34-17-52-26-13-40-20-10-05-16-08-04-02-01-04-02-01-04-02-01-
04-02-01-04-02-01-...

155-466-233-700-350-175-526-263-790-395-1186-593-1780-890-445-1336-668-334-
167-502-251-754-377-1132-566-283-850-425-1276-638-319-958-479-1438-719-2158-
1079-3238-1619-4858-2429-7288-3644-1822-911-2734-1367-4102-2051-6154-3077-
9232-4616-2308-1154-577-1732-866-433-1300-650-325-976-488-244-122-61-184-92-
46-23-70-35-106-53-160-80-40-20-10-05-16-08-04-02--01-04-02-01-04-02-01-04-02-01-
...

157-472-236-118-59-178-89-268-134-67-202-101-304-152-76-38-19-58-29-88-
44-22-11-34-17-52-26-13-40-20-10-05-16-08-04-02-01-04-02-01-04-02-01-
04-02-01-04-02-01-...

159-478-239-718-359-1078-539-1618-809-2428-1214-607-1822-911-2734-1367-
4102-2051-6154-3077-9232-4616-2308-1154-577-1732-866-433-1300-650-325-
976-488-244-122-61-184-92-46-23-70-35-106-53-160-80-40-20-10-05-16-08-04-02-01-
04-02-01-04-02-01-...

161-484-242-121-364-182-91-274-137-412-206-103-310-155-466-233-700-350-175-
526-263-790-395-1186-593-1780-890-445-1336-668-334-167-502-251-754-377-
1132-566-283-850-425-1276-638-319-958-479-1438-719-2158-1079-3238-1619-
4858-2429-7288-3644-1822-911-2734-1367-4102-2051-6154-3077-9232-4616-
2308-1154-577-1732-866-433-1300-650-325-976-488-244-122-61-184-92-46-23-

70-35-106-53-160-80-40-20-10-05-16-08-04-02-01-04-02-01-04-02-01-...

163- 490-245-736-368-184-92-46-23-70-35-106-53-160-80-40-20-10-05-16-08-04-02-
01-04-02-01-04-02-01-04-02-01-
04-02-01-...

165-496-248-124-62-31-94-47-142-71-214-107-322-161-484-242-121-364-182-91-274-
137-412-206-
103-310-155-466-233-700-350-175-526-263-790-395-1186-593-1780-890-445-1336-
668-334-167-502-251-754-377-1132-566-283-850-425-1276-638-319-
958-479-1438-719-2158-1079-3238-1619-4858-2429-7288-3644-1822-911-2734-1367-
4102-2051-6154-3077-9232-4616-2308-1154-577-1732-866-433-1300-
650-325-976-488-244-122-61-184-92-46-23-70-35-106-53-160-80-40-20-10
-05-16-08-04-02-01-04-02-01-04-02-01-04-02-01-04-02-01-...

167-502-251-754-377-1132-566-283-850-425-
1276-638-319-958-479-1438-719-2158-1079-3238-1619-4858-2429-7288-3644-
1822-911-2734-1367-4102-2051-6154-3077-9232-4616-2308-1154-577-1732-
866-433-1300-650-325-976-488-244-122-61-184-92-46-23-70-35-106-53-160-80-40-20-
10-05-16-08-04-02-01-04-02-01-04-02-01-...

169-508-254-127-382-191-574-287-862-431-1294-647-1942-971-2914-1457-4372-2186-
1093-3280-1640-820-410-205-616-308-154-77-232-116-58-29-88-44-22-11-
34-17-52-26-13-40-20-10-05-16-08-04-02-01-04-02-01-04-02-01-...

171-514-257-772-386-193-580-290-145-436-218-109-328-164-82-41-124-62-31-
94-47-142-71-214-107-322-161-484-242-121-364-182-91-274-137-412-206-
103-310-155-466-233-700-350-175-526-263-790-395-1186-593-1780-890-445-
1336-668-334-167-502-251-754-377-1132-566-283-850-425-1276-638-319-958-479-
1438-719-2158-1079-3238-1619-4858-2429-7288-3644-1822-911-2734-
1367-4102-2051-6154-3077-9232-4616-2308-1154-577-1732-866-433-1300-650-325-
976-488-244-122-61-184-92-46-23-70-35-106-53-160-80-40-20-10-05-16-08-04-02-01-
04-02-01-04-02-01-...

173-520-260-130-65-196-98-49-148-74-37-112-56-28-14-07-22-11-34-17-52-
26-13-40-20-10-05-16-08-04-02-01-04-02-01-04-02-01-...

175-526-263-790-395-1186-593-1780-890-445-1336-668-334-167-502-251-
754-377-1132-566-283-850-425-1276-638-319-958-479-1438-719-2158-1079-
3238-1619-4858-2429-7288-3644-1822-911-2734-1367-4102-2051-6154-3077-
9232-4616-2308-1154-577-1732-866-433-1300-650-325-976-488-244-122-61-
184-92-46-23-70-35-106-53-160-80-40-20-10-05-16-08-04-02-01-04-02-01-
04-02-01-04-02-01-...

177-532-266-133-400-200-100-50-25-76-38-19-58-29-88-44-22-11-34-17-52-26-13-40-

20-10-05-16-08-04-02-01-04-02-01-04-02-01-...

179-538-269-808-404-202-101-304-152-76-38-19-58-29-88-44-22-11-34-17-52-26-13-40-20-10-05-16-08-04-02-01-04-02-01-04-02-01-...

181-544-272-136-68-34-17-52-26-13-40-20-10-05-16-08-04-02-01-04-02-01-04-02-01-04-02-01-...

183-550-275-826-413-1240-620-310-155-466-233-700-350-175-526-263-790-395-1186-593-1780-890-445-1336-668-334-167-502-251-754-377-1132-566-283-850-425-1276-638-319-958-479-1438-719-2158-1079-3238-1619-4858-2429-7288-3644-1822-911-2734-1367-4102-2051-6154-3077-9232-4616-2308-1154-577-1732-866-433-1300-650-325-976-488-244-122-61-184-92-46-23-70-35-106-53-160-80-40-20-10-05-16-08-04-02-01-04-02-01-04-02-01-...

185-556-278-139-418-209-628-314-157-472-236-118-59-178-89-268-134-67-202-101-304-152-76-38-19-58-29-88-44-22-11-34-17-52-26-13-40-20-10-05-16-08-04-02-01-04-02-01-04-02-01-...

187-562-281-844-422-211-634-317-952-476-238-119-358-179-538-269-808-404-202-101-304-152-76-38-19-58-29-88-44-22-11-34-17-52-26-13-40-20-10-05-16-08-04-02-01-04-02-01-04-02-01-04-02-01-...

189-568-284-142-71-214-107-322-161-484-242-121-364-182-91-274-137-412-206-103-310-155-466-233-700-350-175-526-263-790-395-1186-593-1780-890-445-1336-668-334-167-502-251-754-377-1132-566-283-850-425-1276-638-319-958-479-1438-719-2158-1079-3238-1619-4858-2429-7288-3644-1822-911-2734-1367-4102-2051-6154-3077-9232-4616-2308-1154-577-1732-866-433-1300-650-325-976-488-244-122-61-184-92-46-23-70-35-106-53-160-80-40-20-10-05-16-08-04-02-01-04-02-01-04-02-01-...

191-574-287-862-431-1294-647-1942-971-2914-1457-4372-2186-1093-3280-1640-820-410-205-616-308-154-77-232-116-58-29-88-44-22-11-34-17-52-26-13-40-20-10-05-16-08-04-02-01-04-02-01-04-02-01-...

193-580-290-145-436-218-109-328-164-82-41-124-62-31-94-47-142-71-214-107-322-161-484-242-121-364-182-91-274-137-412-206-103-310-155-466-233-700-350-175-526-263-790-395-1186-593-1780-890-445-1336-668-334-167-502-251-754-377-1132-566-283-850-425-1276-638-319-958-479-1438-719-2158-1079-3238-1619-4858-2429-7288-3644-1822-911-2734-1367-4102-2051-6154-3077-9232-4616-2308-1154-577-1732-866-433-1300-650-325-976-488-244-122-61-184-92-46-23-70-35-106-53-160-80-40-20-10-05-16-08-04-02-01-04-02-01-

04-02-01-04-02-01-...

195-586-293-880-440-220-110-55-166-83-250-125-376-188-94-47-142-71-214-
107-322-161-484-242-121-364-182-91-274-137-412-206-103-310-155-466-233-
700-350-175-526-263-790-395-1186-593-1780-890-445-1336-668-334-167-502-
251-754-377-1132-566-283-850-425-1276-638-319-958-479-1438-719-2158-
1079-3238-1619-4858-2429-7288-3644-1822-911-2734-1367-4102-2051-6154-
3077-9232-4616-2308-1154-577-1732-866-433-1300-650-325-976-488-244-122-
61-184-92-46-23-70-35-106-53-160-80-40-20-10-05-16-08-04-02-01-04-02-01-
04-02-01-04-02-01-04-02-01-...

197-592-296-148-74-37-112-56-28-14-07-22-11-34-17-52-26-13-40-20-10-05-
16-08-04-02-01-04-02-01-04-02-01-...

199-598-299-898-449-1348-674-337-1012-506-253-760-380-190-95-286-143-
430-215-646-323-970-485-1456-728-364-182-91-274-137-412-206-103-310-
155-466-233-700-350-175-526-263-790-395-1186-593-1780-890-445-1336-668-334-
167-502-251-754-377-1132-566-283-850-425-1276-638-319-958-479-1438-719-2158-
1079-3238-1619-4858-2429-7288-3644-1822-911-2734-1367-4102-2051-6154-3077-
9232-4616-2308-1154-577-1732-866-433-1300-650-325-976-
488-244-122-61-184-92-46-23-70-35-106-53-160-80-40-20-10-05-16-08-04-02-01-04-
02-01-04-02-01-04-02-01-...

201-604-302-151-454-227-682-341-1024-512-256-128-64-32-16-08-04-02-01-04-02-01-
04-02-01-04-02-01-...

203-610-305-916-458-229-688-344-172-86-43-130-65-196-98-49-148-74-37-112-56-28-
14-07-22-11-
34-17-52-26-13-40-20-10-05-16-08-04-02-01-04-02-01-04-02-01-...

205-616-308-154-77-232-116-58-29-88-44-22-11-34-17-52-26-13-40-20-10-05-16-08-
04-02-01-
04-02-01-04-02-01-...

207-622-311-934-467-1402-701-2104-1052-526-263-790-395-1186-593-1780-890-445-
1336-668-334-167-502-251-754-377-1132-566-283-850-425-
1276-638-319-958-479-1438-719-2158-1079-3238-1619-4858-2429-7288-3644-
1822-911-2734-1367-4102-2051-6154-3077-9232-4616-2308-1154-577-1732-
866-433-1300-650-325-976-488-244-122-61-184-92-46-23-70-35-106-53-160-80-40-20-
10-05-16-08-04-02-01-04-02-01-04-02-01-...

209-628-314-157-472-236-118-59-178-89-268-134-67-202-101-304-152-76-38-19-58-
29-88-
44-22-11-34-17-52-26-13-40-20-10-05-16-08-04-02-01-04-02-01-04-02-01-
04-02-01-04-02-01-...

211-634-317-952-476-238-119-358-179-538-269-808-404-202-101-304-152-76-38-19-
58-29-88-44-22-11-

34-17-52-26-13-40-20-10-05-16-08-04-02-01-04-02-01-04-02-01-...

213-640-320-160-80-40-20-10-05-16-08-04-02-01-04-02-01-04-02-01-...

215-646-323-970-485-1456-728-364-182-91-274-137-412-206-103-310-155-466-233-
700-350-175-526-263-790-395-
1186-593-1780-890-445-1336-668-334-167-502-251-754-377-1132-566-283-
850-425-1276-638-319-958-479-1438-719-2158-1079-3238-1619-4858-2429-
7288-3644-1822-911-2734-1367-4102-2051-6154-3077-9232-4616-2308-1154-
577-1732-866-433-1300-650-325-976-488-244-122-61-184-92-46-23-70-35-106-
53-160-80-40-20-10-05-16-08-04-02-01-04-02-01-04-02-01-...

217-652-326-163-490-245-736-368-184-92-46-23-70-35-106-53-160-80-40-20-10-05-
16-08-04-02-01-04-02-01-04-02-01-04-02-01-
04-02-01-...

219-658-329-988-494-247-742-371-1114-557-1672-836-418-209-628-314-157-472-236-
118-59-178-89-268-134-67-202-101-304-

152-76-38-19-58-29-88-44-22-11-34-17-52-26-13-40-20-10-05-16-08-04-02-01-04-02-
01-04-02-01-...

221-664-332-166-83-250-125-376-188-94-47-142-71-214-107-322-161-484-242-
121-364-182-91-274-137-412-206-103-310-155-466-233-700-350-175-526-263-
790-395-1186-593-1780-890-445-1336-668-334-167-502-251-754-377-1132-566-
283-850-425-1276-638-319-958-479-1438-719-2158-1079-3238-1619-4858-2429-
7288-3644-1822-911-2734-1367-4102-2051-6154-3077-9232-4616-2308-1154-
577-1732-866-433-1300-650-325-976-488-244-122-61-184-92-46-23-70-35-106-53-160-
80-40-20-10-05-16-08-04-02-01--04-02-01-04-02-01-...

223-670-335-1006-503-1510-755-2266-1133-3400-1700-850-425-1276-638-319-
958-479-1438-719-2158-1079-3238-1619-4858-2429-7288-3644-1822-911-2734-
1367-4102-2051-6154-3077-9232-4616-2308-1154-577-1732-866-433-1300-650-325-
976-488-244-122-61-184-92-46-23-70-35-106-53-160-80-40-20-10-05-16-08-04-02-01-
04-02-01-04-02-01-...

225-676-338-169-508-254-127-382-191-574-287-862-431-1294-647-1942-971-
2914-1457-4372-2186-1093-3280-1640-820-410-205-616-308-154-77-232-116-
58-29-88-44-22-11-34-17-52-26-13-40-20-10-05-16-08-04-02-01-04-02-01-
04-02-01-04-02-01-04-02-01-...

227-682-341-1024-512-256-128-64-32-16-08-04-02-01-04-02-01-04-02-01-04-
02-01-04-02-01-04-02-01-...

229-688-344-172-86-43-130-65-196-98-49-148-74-37-112-56-28-14-07-22-11-

34-17-52-26-13-40-20-10-05-16-08-04-02-01-04-02-01-04-02-01-...

231-694-347-1042-521-1564-782-391-1174-587-1762-881-2644-1322-661-1984-992-496-248-124-62-31-94-47-142-71-214-107-322-161-484-242-121-364-182-91-274-137-412-206-103-310-155-466-233-700-350-175-526-263-790-395-1186-593-1780-890-445-1336-668-334-167-502-251-754-377-1132-566-283-850-425-1276-638-319-958-479-1438-719-2158-1079-3238-1619-4858-2429-7288-3644-1822-911-2734-1367-4102-2051-6154-3077-9232-4616-2308-1154-577-1732-866-433-1300-650-325-976-488-244-122-61-184-92-46-23-70-35-106-53-160-80-40-20-10-05-16-08-04-02-01-04-02-01-04-02-01-...

```
X(26): (testnumbers)

* * * * * * * * * * * *
1
(EVEN 4 1)
(BRANCHING 2)
(TEST-COMPLETED 1 2)
(BRANCHING 502)
(TEST-COMPLETED 251 502)
* * * * * * * * * * * *
2
(TEST-COMPLETED 1 2)
* * * * * * * * * * * *
3
(EVEN 10 3)
(BRANCHING 5)
(EVEN 16 5)
(BRANCHING 8)
(TEST-COMPLETED 4 8)
(BRANCHING 508)
(TEST-COMPLETED 254 508)
(BRANCHING 505)
(EVEN 516 505)
(BRANCHING 258)
(TEST-COMPLETED 129 258)
(BRANCHING 758)
(TEST-COMPLETED 379 758)
* * * * * * * * * * * *
4
(TEST-COMPLETED 2 4)
* * * * * * * * * * * *
5
(EVEN 16 5)
(BRANCHING 8)
(TEST-COMPLETED 4 8)
(BRANCHING 508)
(TEST-COMPLETED 254 508)
* * * * * * * * * * * *
6
(TEST-COMPLETED 3 6)
* * * * * * * * * * * *
7
(EVEN 22 7)
(BRANCHING 11)
(EVEN 34 11)
(BRANCHING 17)
(EVEN 52 17)
(BRANCHING 26)
(TEST-COMPLETED 13 26)
(BRANCHING 526)
(TEST-COMPLETED 263 526)
(BRANCHING 517)
(EVEN 552 517)
(BRANCHING 276)
(TEST-COMPLETED 138 276)
(BRANCHING 776)
(TEST-COMPLETED 388 776)
```

```
(BRANCHING 511)
(EVEN 534 511)
(BRANCHING 267)
(EVEN 802 267)
(BRANCHING 401)
(TEST-COMPLETED 204 401)
(BRANCHING 901)
(TEST-COMPLETED 704 901)
(BRANCHING 767)
(TEST-COMPLETED 302 767)
************
8
(TEST-COMPLETED 4 8)
************
9
(EVEN 28 9)
(BRANCHING 14)
(TEST-COMPLETED 7 14)
(BRANCHING 514)
(TEST-COMPLETED 257 514)
************
10
(TEST-COMPLETED 5 10)
************
11
(EVEN 34 11)
(BRANCHING 17)
(EVEN 52 17)
(BRANCHING 26)
(TEST-COMPLETED 13 26)
(BRANCHING 526)
(TEST-COMPLETED 263 526)
(BRANCHING 517)
(EVEN 552 517)
(BRANCHING 276)
(TEST-COMPLETED 138 276)
(BRANCHING 776)
(TEST-COMPLETED 388 776)
************
12
(TEST-COMPLETED 6 12)
************
13
(EVEN 40 13)
(BRANCHING 20)
(TEST-COMPLETED 10 20)
(BRANCHING 520)
(TEST-COMPLETED 260 520)
************
14
(TEST-COMPLETED 7 14)
************
15
(EVEN 46 15)
(BRANCHING 23)
(EVEN 70 23)
(BRANCHING 35)
```

```
(EVEN 106 35)
(BRANCHING 53)
(EVEN 160 53)
(BRANCHING 80)
(TEST-COMPLETED 40 80)
(BRANCHING 580)
(TEST-COMPLETED 290 580)
(BRANCHING 553)
(EVEN 660 553)
(BRANCHING 330)
(TEST-COMPLETED 165 330)
(BRANCHING 830)
(TEST-COMPLETED 415 830)
(BRANCHING 535)
(EVEN 606 535)
(BRANCHING 303)
(EVEN 910 303)
(BRANCHING 455)
(TEST-COMPLETED 366 455)
(BRANCHING 955)
(EVEN 866 955)
(BRANCHING 433)
(TEST-COMPLETED 300 433)
(BRANCHING 933)
(TEST-COMPLETED 800 933)
(BRANCHING 803)
(TEST-COMPLETED 410 803)
(BRANCHING 523)
(EVEN 570 523)
(BRANCHING 285)
(EVEN 856 285)
(BRANCHING 428)
(TEST-COMPLETED 214 428)
(BRANCHING 928)
(TEST-COMPLETED 464 928)
(BRANCHING 785)
(TEST-COMPLETED 356 785)
************
16
(TEST-COMPLETED 8 16)
************
17
(EVEN 52 17)
(BRANCHING 26)
(TEST-COMPLETED 13 26)
(BRANCHING 526)
(TEST-COMPLETED 263 526)
************
18
(TEST-COMPLETED 9 18)
************
19
(EVEN 58 19)
(BRANCHING 29)
(EVEN 88 29)
(BRANCHING 44)
(TEST-COMPLETED 22 44)
```

```
(BRANCHING 544)
(TEST-COMPLETED 272 544)
(BRANCHING 529)
(EVEN 588 529)
(BRANCHING 294)
(TEST-COMPLETED 147 294)
(BRANCHING 794)
(TEST-COMPLETED 397 794)
* * * * * * * * * * * *
20
(TEST-COMPLETED 10 20)
* * * * * * * * * * * *
21
(EVEN 64 21)
(BRANCHING 32)
(TEST-COMPLETED 16 32)
(BRANCHING 532)
(TEST-COMPLETED 266 532)
* * * * * * * * * * * *
22
(TEST-COMPLETED 11 22)
* * * * * * * * * * * *
23
(EVEN 70 23)
(BRANCHING 35)
(EVEN 106 35)
(BRANCHING 53)
(EVEN 160 53)
(BRANCHING 80)
(TEST-COMPLETED 40 80)
(BRANCHING 580)
(TEST-COMPLETED 290 580)
(BRANCHING 553)
(EVEN 660 553)
(BRANCHING 330)
(TEST-COMPLETED 165 330)
(BRANCHING 830)
(TEST-COMPLETED 415 830)
(BRANCHING 535)
(EVEN 606 535)
(BRANCHING 303)
(EVEN 910 303)
(BRANCHING 455)
(TEST-COMPLETED 366 455)
(BRANCHING 955)
(EVEN 866 955)

(BRANCHING 433)
(TEST-COMPLETED 300 433)
(BRANCHING 933)
(TEST-COMPLETED 800 933)
(BRANCHING 803)
(TEST-COMPLETED 410 803)
* * * * * * * * * * * *
24
(TEST-COMPLETED 12 24)
* * * * * * * * * * * *
```

```
25
(EVEN 76 25)
(BRANCHING 38)
(TEST-COMPLETED 19 38)
(BRANCHING 538)
(TEST-COMPLETED 269 538)
************
26
(TEST-COMPLETED 13 26)
************
27
(EVEN 82 27)
(BRANCHING 41)
(EVEN 124 41)
(BRANCHING 62)
(TEST-COMPLETED 31 62)
(BRANCHING 562)
(TEST-COMPLETED 281 562)
(BRANCHING 541)
(EVEN 624 541)
(BRANCHING 312)
(TEST-COMPLETED 156 312)
(BRANCHING 812)
(TEST-COMPLETED 406 812)
************
28
(TEST-COMPLETED 14 28)
************
29
(EVEN 88 29)
(BRANCHING 44)
(TEST-COMPLETED 22 44)
(BRANCHING 544)
(TEST-COMPLETED 272 544)
************
30
(TEST-COMPLETED 15 30)
************
31
(EVEN 94 31)
(BRANCHING 47)
(EVEN 142 47)
(BRANCHING 71)
(EVEN 214 71)
(BRANCHING 107)
(EVEN 322 107)
(BRANCHING 161)
(EVEN 484 161)
(BRANCHING 242)
(TEST-COMPLETED 121 242)
(BRANCHING 742)
(TEST-COMPLETED 371 742)
(BRANCHING 661)
(EVEN 984 661)
(BRANCHING 492)
(TEST-COMPLETED 246 492)
(BRANCHING 992)
```

```
(TEST-COMPLETED 496 992)
(BRANCHING 607)
(EVEN 822 607)
(BRANCHING 411)
(TEST-COMPLETED 234 411)
(BRANCHING 911)
(TEST-COMPLETED 734 911)
(BRANCHING 571)
(EVEN 714 571)
(BRANCHING 357)
(TEST-COMPLETED 72 357)
(BRANCHING 857)
(TEST-COMPLETED 572 857)
(BRANCHING 547)
(EVEN 642 547)
(BRANCHING 321)
(EVEN 964 321)
(BRANCHING 482)
(TEST-COMPLETED 241 482)
(BRANCHING 982)
(TEST-COMPLETED 491 982)
(BRANCHING 821)
(TEST-COMPLETED 464 821)
************
32
(TEST-COMPLETED 16 32)
************
33
(EVEN 100 33)
(BRANCHING 50)
(TEST-COMPLETED 25 50)
(BRANCHING 550)
(TEST-COMPLETED 275 550)
************
34
(TEST-COMPLETED 17 34)
************
35
(EVEN 106 35)
(BRANCHING 53)
(EVEN 160 53)
(BRANCHING 80)
(TEST-COMPLETED 40 80)
(BRANCHING 580)
(TEST-COMPLETED 290 580)
(BRANCHING 553)
(EVEN 660 553)
(BRANCHING 330)
(TEST-COMPLETED 165 330)
(BRANCHING 830)
(TEST-COMPLETED 415 830)
************
36
(TEST-COMPLETED 18 36)
************
37
(EVEN 112 37)
```

```
(BRANCHING 56)
(TEST-COMPLETED 28 56)
(BRANCHING 556)
(TEST-COMPLETED 278 556)
************
38
(TEST-COMPLETED 19 38)
************
39
(EVEN 118 39)
(BRANCHING 59)
(EVEN 178 59)
(BRANCHING 89)
(EVEN 268 89)
(BRANCHING 134)
(TEST-COMPLETED 67 134)
(BRANCHING 634)
(TEST-COMPLETED 317 634)
(BRANCHING 589)
(EVEN 768 589)
(BRANCHING 384)
(TEST-COMPLETED 192 384)
(BRANCHING 884)
(TEST-COMPLETED 442 884)
(BRANCHING 559)
(EVEN 678 559)
(BRANCHING 339)
(TEST-COMPLETED 18 339)
(BRANCHING 839)
(TEST-COMPLETED 518 839)
************
40
(TEST-COMPLETED 20 40)
************
41
(EVEN 124 41)
(BRANCHING 62)
(TEST-COMPLETED 31 62)
(BRANCHING 562)
(TEST-COMPLETED 281 562)
************
42
(TEST-COMPLETED 21 42)
************
43
(EVEN 130 43)
(BRANCHING 65)
(EVEN 196 65)
(BRANCHING 98)
(TEST-COMPLETED 49 98)
(BRANCHING 598)
(TEST-COMPLETED 299 598)
(BRANCHING 565)
(EVEN 696 565)
(BRANCHING 348)
(TEST-COMPLETED 174 348)
(BRANCHING 848)
```

```
(TEST-COMPLETED 424 848)
************
44
(TEST-COMPLETED 22 44)
************
45
(EVEN 136 45)
(BRANCHING 68)
(TEST-COMPLETED 34 68)
(BRANCHING 568)
(TEST-COMPLETED 284 568)
************
46
(TEST-COMPLETED 23 46)
************
47
(EVEN 142 47)
(BRANCHING 71)
(EVEN 214 71)
(BRANCHING 107)
(EVEN 322 107)
(BRANCHING 161)
(EVEN 484 161)
(BRANCHING 242)
(TEST-COMPLETED 121 242)
(BRANCHING 742)
(TEST-COMPLETED 371 742)
(BRANCHING 661)
(EVEN 984 661)
(BRANCHING 492)
(TEST-COMPLETED 246 492)
(BRANCHING 992)
(TEST-COMPLETED 496 992)
(BRANCHING 607)
(EVEN 822 607)
(BRANCHING 411)
(TEST-COMPLETED 234 411)
(BRANCHING 911)
(TEST-COMPLETED 734 911)
(BRANCHING 571)
(EVEN 714 571)
(BRANCHING 357)
(TEST-COMPLETED 72 357)
(BRANCHING 857)
(TEST-COMPLETED 572 857)
************
48
(TEST-COMPLETED 24 48)
************
49
(EVEN 148 49)
(BRANCHING 74)
(TEST-COMPLETED 37 74)
(BRANCHING 574)
(TEST-COMPLETED 287 574)
************
50
```

```
(TEST-COMPLETED 25 50)
************
51
(EVEN 154 51)
(BRANCHING 77)
(EVEN 232 77)
(BRANCHING 116)
(TEST-COMPLETED 58 116)
(BRANCHING 616)
(TEST-COMPLETED 308 616)
(BRANCHING 577)
(EVEN 732 577)
(BRANCHING 366)
(TEST-COMPLETED 183 366)
(BRANCHING 866)
(TEST-COMPLETED 433 866)
************
52
(TEST-COMPLETED 26 52)
************
53
(EVEN 160 53)
(BRANCHING 80)
(TEST-COMPLETED 40 80)
(BRANCHING 580)
(TEST-COMPLETED 290 580)
************
54
(TEST-COMPLETED 27 54)
************
55
(EVEN 166 55)
(BRANCHING 83)
(EVEN 250 83)
(BRANCHING 125)
(EVEN 376 125)
(BRANCHING 188)
(TEST-COMPLETED 94 188)
(BRANCHING 688)
(TEST-COMPLETED 344 688)
(BRANCHING 625)
(EVEN 876 625)
(BRANCHING 438)
(TEST-COMPLETED 219 438)
(BRANCHING 938)
(TEST-COMPLETED 469 938)
(BRANCHING 583)
(EVEN 750 583)
(BRANCHING 375)
(TEST-COMPLETED 126 375)
(BRANCHING 875)
(TEST-COMPLETED 626 875)
************
56
(TEST-COMPLETED 28 56)
************
57
```

```
(EVEN 172 57)
(BRANCHING 86)
(TEST-COMPLETED 43 86)
(BRANCHING 586)
(TEST-COMPLETED 293 586)
* * * * * * * * * * * *
58
(TEST-COMPLETED 29 58)
* * * * * * * * * * * *
59
(EVEN 178 59)
(BRANCHING 89)
(EVEN 268 89)
(BRANCHING 134)
(TEST-COMPLETED 67 134)
(BRANCHING 634)
(TEST-COMPLETED 317 634)
(BRANCHING 589)
(EVEN 768 589)
(BRANCHING 384)
(TEST-COMPLETED 192 384)
(BRANCHING 884)
(TEST-COMPLETED 442 884)
* * * * * * * * * * * *
60
(TEST-COMPLETED 30 60)
* * * * * * * * * * * *
61
(EVEN 184 61)
(BRANCHING 92)
(TEST-COMPLETED 46 92)
(BRANCHING 592)
(TEST-COMPLETED 296 592)
* * * * * * * * * * * *
62
(TEST-COMPLETED 31 62)
* * * * * * * * * * * *
63
(EVEN 190 63)
(BRANCHING 95)
(EVEN 286 95)
(BRANCHING 143)
(EVEN 430 143)
(BRANCHING 215)
(EVEN 646 215)
(BRANCHING 323)
(EVEN 970 323)
(BRANCHING 485)
(EVEN 456 485)
(BRANCHING 228)
(TEST-COMPLETED 114 228)
(BRANCHING 728)
(TEST-COMPLETED 364 728)
(BRANCHING 985)
(EVEN 956 985)
(BRANCHING 478)
(TEST-COMPLETED 239 478)
```

```
(BRANCHING 978)
(TEST-COMPLETED 489 978)
(BRANCHING 823)
(TEST-COMPLETED 470 823)
(BRANCHING 715)
(TEST-COMPLETED 146 715)
(BRANCHING 643)
(EVEN 930 643)
(BRANCHING 465)
(TEST-COMPLETED 396 465)
(BRANCHING 965)
(EVEN 896 965)
(BRANCHING 448)
(TEST-COMPLETED 224 448)
(BRANCHING 948)
(TEST-COMPLETED 474 948)
(BRANCHING 595)
(EVEN 786 595)
(BRANCHING 393)
(TEST-COMPLETED 180 393)
(BRANCHING 893)
(TEST-COMPLETED 680 893)
* * * * * * * * * * * *
64
(TEST-COMPLETED 32 64)
* * * * * * * * * * * *
65
(EVEN 196 65)
(BRANCHING 98)
(TEST-COMPLETED 49 98)
(BRANCHING 598)
(TEST-COMPLETED 299 598)
* * * * * * * * * * * *
66
(TEST-COMPLETED 33 66)
* * * * * * * * * * * *
67
(EVEN 202 67)
(BRANCHING 101)
(EVEN 304 101)
(BRANCHING 152)
(TEST-COMPLETED 76 152)
(BRANCHING 652)
(TEST-COMPLETED 326 652)
(BRANCHING 601)
(EVEN 804 601)
(BRANCHING 402)
(TEST-COMPLETED 201 402)
(BRANCHING 902)
(TEST-COMPLETED 451 902)
* * * * * * * * * * * *
68
(TEST-COMPLETED 34 68)
* * * * * * * * * * * *
69
(EVEN 208 69)
(BRANCHING 104)
```

```
(TEST-COMPLETED 52 104)
(BRANCHING 604)
(TEST-COMPLETED 302 604)
************
70
(TEST-COMPLETED 35 70)
************
71
(EVEN 214 71)
(BRANCHING 107)
(EVEN 322 107)
(BRANCHING 161)
(EVEN 484 161)
(BRANCHING 242)
(TEST-COMPLETED 121 242)
(BRANCHING 742)
(TEST-COMPLETED 371 742)
(BRANCHING 661)
(EVEN 984 661)
(BRANCHING 492)
(TEST-COMPLETED 246 492)
(BRANCHING 992)
(TEST-COMPLETED 496 992)
(BRANCHING 607)
(EVEN 822 607)
(BRANCHING 411)
(TEST-COMPLETED 234 411)
(BRANCHING 911)
(TEST-COMPLETED 734 911)
************
72
(TEST-COMPLETED 36 72)
************
73
(EVEN 220 73)
(BRANCHING 110)
(TEST-COMPLETED 55 110)
(BRANCHING 610)
(TEST-COMPLETED 305 610)
************
74
(TEST-COMPLETED 37 74)
************
75
(EVEN 226 75)
(BRANCHING 113)
(EVEN 340 113)
(BRANCHING 170)
(TEST-COMPLETED 85 170)
(BRANCHING 670)
(TEST-COMPLETED 335 670)
(BRANCHING 613)
(EVEN 840 613)
(BRANCHING 420)
(TEST-COMPLETED 210 420)
(BRANCHING 920)
(TEST-COMPLETED 460 920)
```

```
* * * * * * * * * * * *
76
(TEST-COMPLETED 38 76)
* * * * * * * * * * * *
77
(EVEN 232 77)
(BRANCHING 116)
(TEST-COMPLETED 58 116)
(BRANCHING 616)
(TEST-COMPLETED 308 616)
* * * * * * * * * * * *
78
(TEST-COMPLETED 39 78)
* * * * * * * * * * * *
79
(EVEN 238 79)
(BRANCHING 119)
(EVEN 358 119)
(BRANCHING 179)
(EVEN 538 179)
(BRANCHING 269)
(EVEN 808 269)
(BRANCHING 404)
(TEST-COMPLETED 202 404)
(BRANCHING 904)
(TEST-COMPLETED 452 904)
(BRANCHING 769)
(TEST-COMPLETED 308 769)
(BRANCHING 679)
(TEST-COMPLETED 38 679)
(BRANCHING 619)
(EVEN 858 619)
(BRANCHING 429)
(TEST-COMPLETED 288 429)
(BRANCHING 929)
(TEST-COMPLETED 788 929)
* * * * * * * * * * * *
80
(TEST-COMPLETED 40 80)
* * * * * * * * * * * *
81
(EVEN 244 81)
(BRANCHING 122)
(TEST-COMPLETED 61 122)
(BRANCHING 622)
(TEST-COMPLETED 311 622)
* * * * * * * * * * * *
82
(TEST-COMPLETED 41 82)
* * * * * * * * * * * *
83
(EVEN 250 83)
(BRANCHING 125)
(EVEN 376 125)
(BRANCHING 188)
(TEST-COMPLETED 94 188)
(BRANCHING 688)
```

```
(TEST-COMPLETED 344 688)
(BRANCHING 625)
(EVEN 876 625)
(BRANCHING 438)
(TEST-COMPLETED 219 438)
(BRANCHING 938)
(TEST-COMPLETED 469 938)
************
84
(TEST-COMPLETED 42 84)
************
85
(EVEN 256 85)
(BRANCHING 128)
(TEST-COMPLETED 64 128)
(BRANCHING 628)
(TEST-COMPLETED 314 628)
************
86
(TEST-COMPLETED 43 86)
************
87
(EVEN 262 87)
(BRANCHING 131)
(EVEN 394 131)
(BRANCHING 197)
(EVEN 592 197)
(BRANCHING 296)
(TEST-COMPLETED 148 296)
(BRANCHING 796)
(TEST-COMPLETED 398 796)
(BRANCHING 697)
(TEST-COMPLETED 92 697)
(BRANCHING 631)
(EVEN 894 631)
(BRANCHING 447)
(TEST-COMPLETED 342 447)
(BRANCHING 947)
(TEST-COMPLETED 842 947)
************
88
(TEST-COMPLETED 44 88)
************
89
(EVEN 268 89)
(BRANCHING 134)
(TEST-COMPLETED 67 134)
(BRANCHING 634)
(TEST-COMPLETED 317 634)
************
90
(TEST-COMPLETED 45 90)
************
91
(EVEN 274 91)
(BRANCHING 137)
(EVEN 412 137)
```

```
(BRANCHING 206)
(TEST-COMPLETED 103 206)
(BRANCHING 706)
(TEST-COMPLETED 353 706)
(BRANCHING 637)
(EVEN 912 637)
(BRANCHING 456)
(TEST-COMPLETED 228 456)
(BRANCHING 956)
(TEST-COMPLETED 478 956)
************
92
(TEST-COMPLETED 46 92)
************
93
(EVEN 280 93)
(BRANCHING 140)
(TEST-COMPLETED 70 140)
(BRANCHING 640)
(TEST-COMPLETED 320 640)
************
94
(TEST-COMPLETED 47 94)
************
95
(EVEN 286 95)
(BRANCHING 143)
(EVEN 430 143)
(BRANCHING 215)
(EVEN 646 215)
(BRANCHING 323)
(EVEN 970 323)
(BRANCHING 485)
(EVEN 456 485)
(BRANCHING 228)
(TEST-COMPLETED 114 228)
(BRANCHING 728)
(TEST-COMPLETED 364 728)
(BRANCHING 985)
(EVEN 956 985)
(BRANCHING 478)
(TEST-COMPLETED 239 478)
(BRANCHING 978)
(TEST-COMPLETED 489 978)
(BRANCHING 823)
(TEST-COMPLETED 470 823)
(BRANCHING 715)
(TEST-COMPLETED 146 715)
(BRANCHING 643)
(EVEN 930 643)
(BRANCHING 465)
(TEST-COMPLETED 396 465)
(BRANCHING 965)
(EVEN 896 965)
(BRANCHING 448)
(TEST-COMPLETED 224 448)
(BRANCHING 948)
```

```
(TEST-COMPLETED 474 948)
************
96
(TEST-COMPLETED 48 96)
************
97
(EVEN 292 97)
(BRANCHING 146)
(TEST-COMPLETED 73 146)
(BRANCHING 646)
(TEST-COMPLETED 323 646)
************
98
(TEST-COMPLETED 49 98)
************
99
(EVEN 298 99)
(BRANCHING 149)
(EVEN 448 149)
(BRANCHING 224)
(TEST-COMPLETED 112 224)
(BRANCHING 724)
(TEST-COMPLETED 362 724)
(BRANCHING 649)
(EVEN 948 649)
(BRANCHING 474)
(TEST-COMPLETED 237 474)
(BRANCHING 974)
(TEST-COMPLETED 487 974)
************
100
(TEST-COMPLETED 50 100)
************
101
(EVEN 304 101)
(BRANCHING 152)
(TEST-COMPLETED 76 152)
(BRANCHING 652)
(TEST-COMPLETED 326 652)
************
102
(TEST-COMPLETED 51 102)
************
103
(EVEN 310 103)
(BRANCHING 155)
(EVEN 466 155)
(BRANCHING 233)
(EVEN 700 233)
(BRANCHING 350)
(TEST-COMPLETED 175 350)
(BRANCHING 850)
(TEST-COMPLETED 425 850)
(BRANCHING 733)
(TEST-COMPLETED 200 733)
(BRANCHING 655)
(EVEN 966 655)
```

```
(BRANCHING 483)
(EVEN 450 483)
(BRANCHING 225)
(EVEN 676 225)
(BRANCHING 338)
(TEST-COMPLETED 169 338)
(BRANCHING 838)
(TEST-COMPLETED 419 838)
(BRANCHING 725)
(TEST-COMPLETED 176 725)
(BRANCHING 983)
(EVEN 950 983)
(BRANCHING 475)
(TEST-COMPLETED 426 475)
(BRANCHING 975)
(EVEN 926 975)
(BRANCHING 463)
(TEST-COMPLETED 390 463)
(BRANCHING 963)
(EVEN 890 963)
(BRANCHING 445)
(TEST-COMPLETED 336 445)
(BRANCHING 945)
(TEST-COMPLETED 836 945)
************
104
(TEST-COMPLETED 52 104)
************
105
(EVEN 316 105)
(BRANCHING 158)
(TEST-COMPLETED 79 158)
(BRANCHING 658)
(TEST-COMPLETED 329 658)
************
106
(TEST-COMPLETED 53 106)
************
107
(EVEN 322 107)
(BRANCHING 161)
(EVEN 484 161)
(BRANCHING 242)
(TEST-COMPLETED 121 242)
(BRANCHING 742)
(TEST-COMPLETED 371 742)
(BRANCHING 661)
(EVEN 984 661)
(BRANCHING 492)
(TEST-COMPLETED 246 492)
(BRANCHING 992)
(TEST-COMPLETED 496 992)
************
108
(TEST-COMPLETED 54 108)
************
109
```

```
(EVEN 328 109)
(BRANCHING 164)
(TEST-COMPLETED 82 164)
(BRANCHING 664)
(TEST-COMPLETED 332 664)
************
110
(TEST-COMPLETED 55 110)
************
111
(EVEN 334 111)
(BRANCHING 167)
(EVEN 502 167)
(BRANCHING 251)
(EVEN 754 251)
(BRANCHING 377)
(TEST-COMPLETED 132 377)
(BRANCHING 877)
(TEST-COMPLETED 632 877)
(BRANCHING 751)
(TEST-COMPLETED 254 751)
(BRANCHING 667)
(TEST-COMPLETED 2 667)
************
112
(TEST-COMPLETED 56 112)
************
113
(EVEN 340 113)
(BRANCHING 170)
(TEST-COMPLETED 85 170)
(BRANCHING 670)
(TEST-COMPLETED 335 670)
************
114
(TEST-COMPLETED 57 114)
************
115
(EVEN 346 115)
(BRANCHING 173)
(EVEN 520 173)
(BRANCHING 260)
(TEST-COMPLETED 130 260)
(BRANCHING 760)
(TEST-COMPLETED 380 760)
(BRANCHING 673)
(TEST-COMPLETED 20 673)
************
116
(TEST-COMPLETED 58 116)
************
117
(EVEN 352 117)
(BRANCHING 176)
(TEST-COMPLETED 88 176)
(BRANCHING 676)
(TEST-COMPLETED 338 676)
```

```
************
118
(TEST-COMPLETED 59 118)
************
119
(EVEN 358 119)
(BRANCHING 179)
(EVEN 538 179)
(BRANCHING 269)
(EVEN 808 269)
(BRANCHING 404)
(TEST-COMPLETED 202 404)
(BRANCHING 904)
(TEST-COMPLETED 452 904)
(BRANCHING 769)
(TEST-COMPLETED 308 769)
(BRANCHING 679)
(TEST-COMPLETED 38 679)
************
120
(TEST-COMPLETED 60 120)
************
121
(EVEN 364 121)
(BRANCHING 182)
(TEST-COMPLETED 91 182)
(BRANCHING 682)
(TEST-COMPLETED 341 682)
************
122
(TEST-COMPLETED 61 122)
************
123
(EVEN 370 123)
(BRANCHING 185)
(EVEN 556 185)
(BRANCHING 278)
(TEST-COMPLETED 139 278)
(BRANCHING 778)
(TEST-COMPLETED 389 778)
(BRANCHING 685)
(TEST-COMPLETED 56 685)
************
124
(TEST-COMPLETED 62 124)
************
125
(EVEN 376 125)
(BRANCHING 188)
(TEST-COMPLETED 94 188)
(BRANCHING 688)
(TEST-COMPLETED 344 688)
************
126
(TEST-COMPLETED 63 126)
************
127
```

```
(EVEN 382 127)
(BRANCHING 191)
(EVEN 574 191)
(BRANCHING 287)
(EVEN 862 287)
(BRANCHING 431)
(TEST-COMPLETED 294 431)
(BRANCHING 931)
(TEST-COMPLETED 794 931)
(BRANCHING 787)
(TEST-COMPLETED 362 787)
(BRANCHING 691)
(TEST-COMPLETED 74 691)
************
128
(TEST-COMPLETED 64 128)
************
129
(EVEN 388 129)
(BRANCHING 194)
(TEST-COMPLETED 97 194)
(BRANCHING 694)
(TEST-COMPLETED 347 694)
************
130
(TEST-COMPLETED 65 130)
************
131
(EVEN 394 131)
(BRANCHING 197)
(EVEN 592 197)
(BRANCHING 296)
(TEST-COMPLETED 148 296)
(BRANCHING 796)
(TEST-COMPLETED 398 796)
(BRANCHING 697)
(TEST-COMPLETED 92 697)
************
132
(TEST-COMPLETED 66 132)
************
133
(EVEN 400 133)
(BRANCHING 200)
(TEST-COMPLETED 100 200)
(BRANCHING 700)
(TEST-COMPLETED 350 700)
************
134
(TEST-COMPLETED 67 134)
************
135
(EVEN 406 135)
(BRANCHING 203)
(EVEN 610 203)
(BRANCHING 305)
(EVEN 916 305)
```

```
(BRANCHING 458)
(TEST-COMPLETED 229 458)
(BRANCHING 958)
(TEST-COMPLETED 479 958)
(BRANCHING 805)
(TEST-COMPLETED 416 805)
(BRANCHING 703)
(TEST-COMPLETED 110 703)
* * * * * * * * * * * *
136
(TEST-COMPLETED 68 136)
* * * * * * * * * * * *
137
(EVEN 412 137)
(BRANCHING 206)
(TEST-COMPLETED 103 206)
(BRANCHING 706)
(TEST-COMPLETED 353 706)
* * * * * * * * * * * *
138
(TEST-COMPLETED 69 138)
* * * * * * * * * * * *
139
(EVEN 418 139)
(BRANCHING 209)
(EVEN 628 209)
(BRANCHING 314)
(TEST-COMPLETED 157 314)
(BRANCHING 814)
(TEST-COMPLETED 407 814)
(BRANCHING 709)
(TEST-COMPLETED 128 709)
* * * * * * * * * * * *
140
(TEST-COMPLETED 70 140)
* * * * * * * * * * * *
141
(EVEN 424 141)
(BRANCHING 212)
(TEST-COMPLETED 106 212)
(BRANCHING 712)
(TEST-COMPLETED 356 712)
* * * * * * * * * * * *
142
(TEST-COMPLETED 71 142)
* * * * * * * * * * * *
143
(EVEN 430 143)
(BRANCHING 215)
(EVEN 646 215)
(BRANCHING 323)
(EVEN 970 323)
(BRANCHING 485)
(EVEN 456 485)
(BRANCHING 228)
(TEST-COMPLETED 114 228)
(BRANCHING 728)
```

```
(TEST-COMPLETED 364 728)
(BRANCHING 985)
(EVEN 956 985)
(BRANCHING 478)
(TEST-COMPLETED 239 478)
(BRANCHING 978)
(TEST-COMPLETED 489 978)
(BRANCHING 823)
(TEST-COMPLETED 470 823)
(BRANCHING 715)
(TEST-COMPLETED 146 715)
************
144
(TEST-COMPLETED 72 144)
************
145
(EVEN 436 145)
(BRANCHING 218)
(TEST-COMPLETED 109 218)
(BRANCHING 718)
(TEST-COMPLETED 359 718)
************
146
(TEST-COMPLETED 73 146)
************
147
(EVEN 442 147)
(BRANCHING 221)
(EVEN 664 221)
(BRANCHING 332)
(TEST-COMPLETED 166 332)
(BRANCHING 832)
(TEST-COMPLETED 416 832)
(BRANCHING 721)
(TEST-COMPLETED 164 721)
************
148
(TEST-COMPLETED 74 148)
************
149
(EVEN 448 149)
(BRANCHING 224)
(TEST-COMPLETED 112 224)
(BRANCHING 724)
(TEST-COMPLETED 362 724)
************
150
(TEST-COMPLETED 75 150)
************
151
(EVEN 454 151)
(BRANCHING 227)
(EVEN 682 227)
(BRANCHING 341)
(TEST-COMPLETED 24 341)
(BRANCHING 841)
(TEST-COMPLETED 524 841)
```

```
(BRANCHING 727)
(TEST-COMPLETED 182 727)
************
152
(TEST-COMPLETED 76 152)
************
153
(EVEN 460 153)
(BRANCHING 230)
(TEST-COMPLETED 115 230)
(BRANCHING 730)
(TEST-COMPLETED 365 730)
************
154
(TEST-COMPLETED 77 154)
************
155
(EVEN 466 155)
(BRANCHING 233)
(EVEN 700 233)
(BRANCHING 350)
(TEST-COMPLETED 175 350)
(BRANCHING 850)
(TEST-COMPLETED 425 850)
(BRANCHING 733)
(TEST-COMPLETED 200 733)
************
156
(TEST-COMPLETED 78 156)
************
157
(EVEN 472 157)
(BRANCHING 236)
(TEST-COMPLETED 118 236)
(BRANCHING 736)
(TEST-COMPLETED 368 736)
************
158
(TEST-COMPLETED 79 158)
************
159
(EVEN 478 159)
(BRANCHING 239)
(EVEN 718 239)
(BRANCHING 359)
(TEST-COMPLETED 78 359)
(BRANCHING 859)
(TEST-COMPLETED 578 859)
(BRANCHING 739)
(TEST-COMPLETED 218 739)
************
160
(TEST-COMPLETED 80 160)
************
161
(EVEN 484 161)
(BRANCHING 242)
```

```
(TEST-COMPLETED 121 242)
(BRANCHING 742)
(TEST-COMPLETED 371 742)
************
162
(TEST-COMPLETED 81 162)
************
163
(EVEN 490 163)
(BRANCHING 245)
(EVEN 736 245)
(BRANCHING 368)
(TEST-COMPLETED 184 368)
(BRANCHING 868)
(TEST-COMPLETED 434 868)
(BRANCHING 745)
(TEST-COMPLETED 236 745)
************
164
(TEST-COMPLETED 82 164)
************
165
(EVEN 496 165)
(BRANCHING 248)
(TEST-COMPLETED 124 248)
(BRANCHING 748)
(TEST-COMPLETED 374 748)
************
166
(TEST-COMPLETED 83 166)
************
167
(EVEN 502 167)
(BRANCHING 251)
(EVEN 754 251)
(BRANCHING 377)
(TEST-COMPLETED 132 377)
(BRANCHING 877)
(TEST-COMPLETED 632 877)
(BRANCHING 751)
(TEST-COMPLETED 254 751)
************
168
(TEST-COMPLETED 84 168)
************
169
(EVEN 508 169)
(BRANCHING 254)
(TEST-COMPLETED 127 254)
(BRANCHING 754)
(TEST-COMPLETED 377 754)
************
170
(TEST-COMPLETED 85 170)
************
171
(EVEN 514 171)
```

```
(BRANCHING 257)
(EVEN 772 257)
(BRANCHING 386)
(TEST-COMPLETED 193 386)
(BRANCHING 886)
(TEST-COMPLETED 443 886)
(BRANCHING 757)
(TEST-COMPLETED 272 757)
* * * * * * * * * * * *
172
(TEST-COMPLETED 86 172)
* * * * * * * * * * * *
173
(EVEN 520 173)
(BRANCHING 260)
(TEST-COMPLETED 130 260)
(BRANCHING 760)
(TEST-COMPLETED 380 760)
* * * * * * * * * * * *
174
(TEST-COMPLETED 87 174)
* * * * * * * * * * * *
175
(EVEN 526 175)
(BRANCHING 263)
(EVEN 790 263)
(BRANCHING 395)
(TEST-COMPLETED 186 395)
(BRANCHING 895)
(TEST-COMPLETED 686 895)
(BRANCHING 763)
(TEST-COMPLETED 290 763)
* * * * * * * * * * * *
176
(TEST-COMPLETED 88 176)
* * * * * * * * * * * *
177
(EVEN 532 177)
(BRANCHING 266)
(TEST-COMPLETED 133 266)
(BRANCHING 766)
(TEST-COMPLETED 383 766)
* * * * * * * * * * * *
178
(TEST-COMPLETED 89 178)
* * * * * * * * * * * *
179
(EVEN 538 179)
(BRANCHING 269)
(EVEN 808 269)
(BRANCHING 404)
(TEST-COMPLETED 202 404)
(BRANCHING 904)
(TEST-COMPLETED 452 904)
(BRANCHING 769)
(TEST-COMPLETED 308 769)
* * * * * * * * * * * *
```

```
180
(TEST-COMPLETED 90 180)
************
181
(EVEN 544 181)
(BRANCHING 272)
(TEST-COMPLETED 136 272)
(BRANCHING 772)
(TEST-COMPLETED 386 772)
************
182
(TEST-COMPLETED 91 182)
************
183
(EVEN 550 183)
(BRANCHING 275)
(EVEN 826 275)
(BRANCHING 413)
(TEST-COMPLETED 240 413)
(BRANCHING 913)
(TEST-COMPLETED 740 913)
(BRANCHING 775)
(TEST-COMPLETED 326 775)
************
184
(TEST-COMPLETED 92 184)
************
185
(EVEN 556 185)
(BRANCHING 278)
(TEST-COMPLETED 139 278)
(BRANCHING 778)
(TEST-COMPLETED 389 778)
************
186
(TEST-COMPLETED 93 186)
************
187
(EVEN 562 187)
(BRANCHING 281)
(EVEN 844 281)
(BRANCHING 422)
(TEST-COMPLETED 211 422)
(BRANCHING 922)
(TEST-COMPLETED 461 922)
(BRANCHING 781)
(TEST-COMPLETED 344 781)
************
188
(TEST-COMPLETED 94 188)
************
189
(EVEN 568 189)
(BRANCHING 284)
(TEST-COMPLETED 142 284)
(BRANCHING 784)
(TEST-COMPLETED 392 784)
```

```
************
190
(TEST-COMPLETED 95 190)
************
191
(EVEN 574 191)
(BRANCHING 287)
(EVEN 862 287)
(BRANCHING 431)
(TEST-COMPLETED 294 431)
(BRANCHING 931)
(TEST-COMPLETED 794 931)
(BRANCHING 787)
(TEST-COMPLETED 362 787)
************
192
(TEST-COMPLETED 96 192)
************
193
(EVEN 580 193)
(BRANCHING 290)
(TEST-COMPLETED 145 290)
(BRANCHING 790)
(TEST-COMPLETED 395 790)
************
194
(TEST-COMPLETED 97 194)
************
195
(EVEN 586 195)
(BRANCHING 293)
(EVEN 880 293)
(BRANCHING 440)
(TEST-COMPLETED 220 440)
(BRANCHING 940)
(TEST-COMPLETED 470 940)
(BRANCHING 793)
(TEST-COMPLETED 380 793)
************
196
(TEST-COMPLETED 98 196)
************
197
(EVEN 592 197)
(BRANCHING 296)
(TEST-COMPLETED 148 296)
(BRANCHING 796)
(TEST-COMPLETED 398 796)
************
198
(TEST-COMPLETED 99 198)
************
199
(EVEN 598 199)
(BRANCHING 299)
(EVEN 898 299)
(BRANCHING 449)
```

```
(TEST-COMPLETED 348 449)
(BRANCHING 949)
(TEST-COMPLETED 848 949)
(BRANCHING 799)
(TEST-COMPLETED 398 799)
* * * * * * * * * * * *
200
(TEST-COMPLETED 100 200)
* * * * * * * * * * * *
201
(EVEN 604 201)
(BRANCHING 302)
(TEST-COMPLETED 151 302)
(BRANCHING 802)
(TEST-COMPLETED 401 802)
* * * * * * * * * * * *
202
(TEST-COMPLETED 101 202)
* * * * * * * * * * * *
203
(EVEN 610 203)
(BRANCHING 305)
(EVEN 916 305)
(BRANCHING 458)
(TEST-COMPLETED 229 458)
(BRANCHING 958)
(TEST-COMPLETED 479 958)
(BRANCHING 805)
(TEST-COMPLETED 416 805)
* * * * * * * * * * * *
204
(TEST-COMPLETED 102 204)
* * * * * * * * * * * *
205
(EVEN 616 205)
(BRANCHING 308)
(TEST-COMPLETED 154 308)
(BRANCHING 808)
(TEST-COMPLETED 404 808)
* * * * * * * * * * * *
206
(TEST-COMPLETED 103 206)
* * * * * * * * * * * *
207
(EVEN 622 207)
(BRANCHING 311)
(EVEN 934 311)
(BRANCHING 467)
(TEST-COMPLETED 402 467)
(BRANCHING 967)
(EVEN 902 967)
(BRANCHING 451)
(TEST-COMPLETED 354 451)
(BRANCHING 951)
(TEST-COMPLETED 854 951)
(BRANCHING 811)
(TEST-COMPLETED 434 811)
```

```
************
208
(TEST-COMPLETED 104 208)
************
209
(EVEN 628 209)
(BRANCHING 314)
(TEST-COMPLETED 157 314)
(BRANCHING 814)
(TEST-COMPLETED 407 814)
************
210
(TEST-COMPLETED 105 210)
************
211
(EVEN 634 211)
(BRANCHING 317)
(EVEN 952 317)
(BRANCHING 476)
(TEST-COMPLETED 238 476)
(BRANCHING 976)
(TEST-COMPLETED 488 976)
(BRANCHING 817)
(TEST-COMPLETED 452 817)
************
212
(TEST-COMPLETED 106 212)
************
213
(EVEN 640 213)
(BRANCHING 320)
(TEST-COMPLETED 160 320)
(BRANCHING 820)
(TEST-COMPLETED 410 820)
************
214
(TEST-COMPLETED 107 214)
************
215
(EVEN 646 215)
(BRANCHING 323)
(EVEN 970 323)
(BRANCHING 485)
(EVEN 456 485)
(BRANCHING 228)
(TEST-COMPLETED 114 228)
(BRANCHING 728)
(TEST-COMPLETED 364 728)
(BRANCHING 985)
(EVEN 956 985)
(BRANCHING 478)
(TEST-COMPLETED 239 478)
(BRANCHING 978)
(TEST-COMPLETED 489 978)
(BRANCHING 823)
(TEST-COMPLETED 470 823)
************
```

```
216
(TEST-COMPLETED 108 216)
************
217
(EVEN 652 217)
(BRANCHING 326)
(TEST-COMPLETED 163 326)
(BRANCHING 826)
(TEST-COMPLETED 413 826)
************
218
(TEST-COMPLETED 109 218)
************
219
(EVEN 658 219)
(BRANCHING 329)
(EVEN 988 329)
(BRANCHING 494)
(TEST-COMPLETED 247 494)
(BRANCHING 994)
(TEST-COMPLETED 497 994)
(BRANCHING 829)
(TEST-COMPLETED 488 829)
************
220
(TEST-COMPLETED 110 220)
************
221
(EVEN 664 221)
(BRANCHING 332)
(TEST-COMPLETED 166 332)
(BRANCHING 832)
(TEST-COMPLETED 416 832)
************
222
(TEST-COMPLETED 111 222)
************
223
(EVEN 670 223)
(BRANCHING 335)
(TEST-COMPLETED 6 335)
(BRANCHING 835)
(TEST-COMPLETED 506 835)
************
224
(TEST-COMPLETED 112 224)
************
225
(EVEN 676 225)
(BRANCHING 338)
(TEST-COMPLETED 169 338)
(BRANCHING 838)
(TEST-COMPLETED 419 838)
************
226
(TEST-COMPLETED 113 226)
************
```

```
227
(EVEN 682 227)
(BRANCHING 341)
(TEST-COMPLETED 24 341)
(BRANCHING 841)
(TEST-COMPLETED 524 841)
* * * * * * * * * * * *
228
(TEST-COMPLETED 114 228)
* * * * * * * * * * * *
229
(EVEN 688 229)
(BRANCHING 344)
(TEST-COMPLETED 172 344)
(BRANCHING 844)
(TEST-COMPLETED 422 844)
* * * * * * * * * * * *
230
(TEST-COMPLETED 115 230)
* * * * * * * * * * * *
231
(EVEN 694 231)
(BRANCHING 347)
(TEST-COMPLETED 42 347)
(BRANCHING 847)
(TEST-COMPLETED 542 847)
* * * * * * * * * * * *
232
(TEST-COMPLETED 116 232)
* * * * * * * * * * * *
233
(EVEN 700 233)
(BRANCHING 350)
(TEST-COMPLETED 175 350)
(BRANCHING 850)
(TEST-COMPLETED 425 850)
* * * * * * * * * * * *
234
(TEST-COMPLETED 117 234)
* * * * * * * * * * * *
235
(EVEN 706 235)
(BRANCHING 353)
(TEST-COMPLETED 60 353)
(BRANCHING 853)
(TEST-COMPLETED 560 853)
* * * * * * * * * * * *
236
(TEST-COMPLETED 118 236)
* * * * * * * * * * * *
237
(EVEN 712 237)
(BRANCHING 356)
(TEST-COMPLETED 178 356)
(BRANCHING 856)
(TEST-COMPLETED 428 856)
* * * * * * * * * * * *
```

```
238
(TEST-COMPLETED 119 238)
************
239
(EVEN 718 239)
(BRANCHING 359)
(TEST-COMPLETED 78 359)
(BRANCHING 859)
(TEST-COMPLETED 578 859)
************
240
(TEST-COMPLETED 120 240)
************
241
(EVEN 724 241)
(BRANCHING 362)
(TEST-COMPLETED 181 362)
(BRANCHING 862)
(TEST-COMPLETED 431 862)
************
242
(TEST-COMPLETED 121 242)
************
243
(EVEN 730 243)
(BRANCHING 365)
(TEST-COMPLETED 96 365)
(BRANCHING 865)
(TEST-COMPLETED 596 865)
************
244
(TEST-COMPLETED 122 244)
************
245
(EVEN 736 245)
(BRANCHING 368)
(TEST-COMPLETED 184 368)
(BRANCHING 868)
(TEST-COMPLETED 434 868)
************
246
(TEST-COMPLETED 123 246)
************
247
(EVEN 742 247)
(BRANCHING 371)
(TEST-COMPLETED 114 371)
(BRANCHING 871)
(TEST-COMPLETED 614 871)
************
248
(TEST-COMPLETED 124 248)
************
249
(EVEN 748 249)
(BRANCHING 374)
(TEST-COMPLETED 187 374)
```

```
(BRANCHING 874)
(TEST-COMPLETED 437 874)
************
250
(TEST-COMPLETED 125 250)
************
251
(EVEN 754 251)
(BRANCHING 377)
(TEST-COMPLETED 132 377)
(BRANCHING 877)
(TEST-COMPLETED 632 877)
************
252
(TEST-COMPLETED 126 252)
************
253
(EVEN 760 253)
(BRANCHING 380)
(TEST-COMPLETED 190 380)
(BRANCHING 880)
(TEST-COMPLETED 440 880)
************
254
(TEST-COMPLETED 127 254)
************
255
(EVEN 766 255)
(BRANCHING 383)
(TEST-COMPLETED 150 383)
(BRANCHING 883)
(TEST-COMPLETED 650 883)
************
256
(TEST-COMPLETED 128 256)
************
257
(EVEN 772 257)
(BRANCHING 386)
(TEST-COMPLETED 193 386)
(BRANCHING 886)
(TEST-COMPLETED 443 886)
************
258
(TEST-COMPLETED 129 258)
************
259
(EVEN 778 259)
(BRANCHING 389)
(TEST-COMPLETED 168 389)
(BRANCHING 889)
(TEST-COMPLETED 668 889)
************
260
(TEST-COMPLETED 130 260)
************
261
```

```
(EVEN 784 261)
(BRANCHING 392)
(TEST-COMPLETED 196 392)
(BRANCHING 892)
(TEST-COMPLETED 446 892)
* * * * * * * * * * * *
262
(TEST-COMPLETED 131 262)
* * * * * * * * * * * *
263
(EVEN 790 263)
(BRANCHING 395)
(TEST-COMPLETED 186 395)
(BRANCHING 895)
(TEST-COMPLETED 686 895)
* * * * * * * * * * * *
264
(TEST-COMPLETED 132 264)
* * * * * * * * * * * *
265
(EVEN 796 265)
(BRANCHING 398)
(TEST-COMPLETED 199 398)
(BRANCHING 898)
(TEST-COMPLETED 449 898)
* * * * * * * * * * * *
266
(TEST-COMPLETED 133 266)
* * * * * * * * * * * *
267
(EVEN 802 267)
(BRANCHING 401)
(TEST-COMPLETED 204 401)
(BRANCHING 901)
(TEST-COMPLETED 704 901)
* * * * * * * * * * * *
268
(TEST-COMPLETED 134 268)
* * * * * * * * * * * *
269
(EVEN 808 269)
(BRANCHING 404)
(TEST-COMPLETED 202 404)
(BRANCHING 904)
(TEST-COMPLETED 452 904)
* * * * * * * * * * * *
270
(TEST-COMPLETED 135 270)
* * * * * * * * * * * *
271
(EVEN 814 271)
(BRANCHING 407)
(TEST-COMPLETED 222 407)
(BRANCHING 907)
(TEST-COMPLETED 722 907)
* * * * * * * * * * * *
272
```

```
(TEST-COMPLETED 136 272)
************
273
(EVEN 820 273)
(BRANCHING 410)
(TEST-COMPLETED 205 410)
(BRANCHING 910)
(TEST-COMPLETED 455 910)
************
274
(TEST-COMPLETED 137 274)
************
275
(EVEN 826 275)
(BRANCHING 413)
(TEST-COMPLETED 240 413)
(BRANCHING 913)
(TEST-COMPLETED 740 913)
************
276
(TEST-COMPLETED 138 276)
************
277
(EVEN 832 277)
(BRANCHING 416)
(TEST-COMPLETED 208 416)
(BRANCHING 916)
(TEST-COMPLETED 458 916)
************
278
(TEST-COMPLETED 139 278)
************
279
(EVEN 838 279)
(BRANCHING 419)
(TEST-COMPLETED 258 419)
(BRANCHING 919)
(TEST-COMPLETED 758 919)
************
280
(TEST-COMPLETED 140 280)
************
281
(EVEN 844 281)
(BRANCHING 422)
(TEST-COMPLETED 211 422)
(BRANCHING 922)
(TEST-COMPLETED 461 922)
************
282
(TEST-COMPLETED 141 282)
************
283
(EVEN 850 283)
(BRANCHING 425)
(TEST-COMPLETED 276 425)
(BRANCHING 925)
```

```
(TEST-COMPLETED 776 925)
************
284
(TEST-COMPLETED 142 284)
************
285
(EVEN 856 285)
(BRANCHING 428)
(TEST-COMPLETED 214 428)
(BRANCHING 928)
(TEST-COMPLETED 464 928)
************
286
(TEST-COMPLETED 143 286)
************
287
(EVEN 862 287)
(BRANCHING 431)
(TEST-COMPLETED 294 431)
(BRANCHING 931)
(TEST-COMPLETED 794 931)
************
288
(TEST-COMPLETED 144 288)
************
289
(EVEN 868 289)
(BRANCHING 434)
(TEST-COMPLETED 217 434)
(BRANCHING 934)
(TEST-COMPLETED 467 934)
************
290
(TEST-COMPLETED 145 290)
************
291
(EVEN 874 291)
(BRANCHING 437)
(TEST-COMPLETED 312 437)
(BRANCHING 937)
(TEST-COMPLETED 812 937)
************
292
(TEST-COMPLETED 146 292)
************
293
(EVEN 880 293)
(BRANCHING 440)
(TEST-COMPLETED 220 440)
(BRANCHING 940)
(TEST-COMPLETED 470 940)
************
294
(TEST-COMPLETED 147 294)
************
295
(EVEN 886 295)
```

```
(BRANCHING 443)
(TEST-COMPLETED 330 443)
(BRANCHING 943)
(TEST-COMPLETED 830 943)
* * * * * * * * * * * *
296
(TEST-COMPLETED 148 296)
* * * * * * * * * * * *
297
(EVEN 892 297)
(BRANCHING 446)
(TEST-COMPLETED 223 446)
(BRANCHING 946)
(TEST-COMPLETED 473 946)
* * * * * * * * * * * *
298
(TEST-COMPLETED 149 298)
* * * * * * * * * * * *
299
(EVEN 898 299)
(BRANCHING 449)
(TEST-COMPLETED 348 449)
(BRANCHING 949)
(TEST-COMPLETED 848 949)
* * * * * * * * * * * *
300
(TEST-COMPLETED 150 300)
* * * * * * * * * * * *
301
(EVEN 904 301)
(BRANCHING 452)
(TEST-COMPLETED 226 452)
(BRANCHING 952)
(TEST-COMPLETED 476 952)
* * * * * * * * * * * *
302
(TEST-COMPLETED 151 302)
* * * * * * * * * * * *
303
(EVEN 910 303)
(BRANCHING 455)
(TEST-COMPLETED 366 455)
(BRANCHING 955)
(EVEN 866 955)
(BRANCHING 433)
(TEST-COMPLETED 300 433)
(BRANCHING 933)
(TEST-COMPLETED 800 933)
* * * * * * * * * * * *
304
(TEST-COMPLETED 152 304)
* * * * * * * * * * * *
305
(EVEN 916 305)
(BRANCHING 458)
(TEST-COMPLETED 229 458)
(BRANCHING 958)
```

```
(TEST-COMPLETED 479 958)
************
306
(TEST-COMPLETED 153 306)
************
307
(EVEN 922 307)
(BRANCHING 461)
(TEST-COMPLETED 384 461)
(BRANCHING 961)
(EVEN 884 961)
(BRANCHING 442)
(TEST-COMPLETED 221 442)
(BRANCHING 942)
(TEST-COMPLETED 471 942)
************
308
(TEST-COMPLETED 154 308)
************
309
(EVEN 928 309)
(BRANCHING 464)
(TEST-COMPLETED 232 464)
(BRANCHING 964)
(TEST-COMPLETED 482 964)
************
310
(TEST-COMPLETED 155 310)
************
311
(EVEN 934 311)
(BRANCHING 467)
(TEST-COMPLETED 402 467)
(BRANCHING 967)
(EVEN 902 967)
(BRANCHING 451)
(TEST-COMPLETED 354 451)
(BRANCHING 951)
(TEST-COMPLETED 854 951)
************
312
(TEST-COMPLETED 156 312)
************
313
(EVEN 940 313)
(BRANCHING 470)
(TEST-COMPLETED 235 470)
(BRANCHING 970)
(TEST-COMPLETED 485 970)
************
314
(TEST-COMPLETED 157 314)
************
315
(EVEN 946 315)
(BRANCHING 473)
(TEST-COMPLETED 420 473)
```

```
(BRANCHING 973)
(EVEN 920 973)
(BRANCHING 460)
(TEST-COMPLETED 230 460)
(BRANCHING 960)
(TEST-COMPLETED 480 960)
* * * * * * * * * * * *
316
(TEST-COMPLETED 158 316)
* * * * * * * * * * * *
317
(EVEN 952 317)
(BRANCHING 476)
(TEST-COMPLETED 238 476)
(BRANCHING 976)
(TEST-COMPLETED 488 976)
* * * * * * * * * * * *
318
(TEST-COMPLETED 159 318)
* * * * * * * * * * * *
319
(EVEN 958 319)
(BRANCHING 479)
(EVEN 438 479)
(BRANCHING 219)
(EVEN 658 219)
(BRANCHING 329)
(EVEN 988 329)
(BRANCHING 494)
(TEST-COMPLETED 247 494)
(BRANCHING 994)
(TEST-COMPLETED 497 994)
(BRANCHING 829)
(TEST-COMPLETED 488 829)
(BRANCHING 719)
(TEST-COMPLETED 158 719)
(BRANCHING 979)
(EVEN 938 979)
(BRANCHING 469)
(TEST-COMPLETED 408 469)
(BRANCHING 969)
(EVEN 908 969)
(BRANCHING 454)
(TEST-COMPLETED 227 454)
(BRANCHING 954)
(TEST-COMPLETED 477 954)
* * * * * * * * * * * *
320
(TEST-COMPLETED 160 320)
* * * * * * * * * * * *
321
(EVEN 964 321)
(BRANCHING 482)
(TEST-COMPLETED 241 482)
(BRANCHING 982)
(TEST-COMPLETED 491 982)
* * * * * * * * * * * *
```

```
322
(TEST-COMPLETED 161 322)
************
323
(EVEN 970 323)
(BRANCHING 485)
(EVEN 456 485)
(BRANCHING 228)
(TEST-COMPLETED 114 228)
(BRANCHING 728)
(TEST-COMPLETED 364 728)
(BRANCHING 985)
(EVEN 956 985)
(BRANCHING 478)
(TEST-COMPLETED 239 478)
(BRANCHING 978)
(TEST-COMPLETED 489 978)
************
324
(TEST-COMPLETED 162 324)
************
325
(EVEN 976 325)
(BRANCHING 488)
(TEST-COMPLETED 244 488)
(BRANCHING 988)
(TEST-COMPLETED 494 988)
************
326
(TEST-COMPLETED 163 326)
************
327
(EVEN 982 327)
(BRANCHING 491)
(EVEN 474 491)
(BRANCHING 237)
(EVEN 712 237)
(BRANCHING 356)
(TEST-COMPLETED 178 356)
(BRANCHING 856)
(TEST-COMPLETED 428 856)
(BRANCHING 737)
(TEST-COMPLETED 212 737)
(BRANCHING 991)
(EVEN 974 991)
(BRANCHING 487)
(EVEN 462 487)
(BRANCHING 231)
(EVEN 694 231)
(BRANCHING 347)
(TEST-COMPLETED 42 347)
(BRANCHING 847)
(TEST-COMPLETED 542 847)
(BRANCHING 731)
(TEST-COMPLETED 194 731)
(BRANCHING 987)
(EVEN 962 987)
```

```
(BRANCHING 481)
(EVEN 444 481)
(BRANCHING 222)
(TEST-COMPLETED 111 222)
(BRANCHING 722)
(TEST-COMPLETED 361 722)
(BRANCHING 981)
(EVEN 944 981)
(BRANCHING 472)
(TEST-COMPLETED 236 472)
(BRANCHING 972)
(TEST-COMPLETED 486 972)
************
328
(TEST-COMPLETED 164 328)
************
329
(EVEN 988 329)
(BRANCHING 494)
(TEST-COMPLETED 247 494)
(BRANCHING 994)
(TEST-COMPLETED 497 994)
************
330
(TEST-COMPLETED 165 330)
************
331
(EVEN 994 331)
(BRANCHING 497)
(EVEN 492 497)
(BRANCHING 246)
(TEST-COMPLETED 123 246)
(BRANCHING 746)
(TEST-COMPLETED 373 746)
(BRANCHING 997)
(EVEN 992 997)
(BRANCHING 496)
(TEST-COMPLETED 248 496)
(BRANCHING 996)
(TEST-COMPLETED 498 996)
************
332
(TEST-COMPLETED 166 332)
************
333
(TEST-COMPLETED 0 333)
************
334
(TEST-COMPLETED 167 334)
************
335
(TEST-COMPLETED 6 335)
************
336
(TEST-COMPLETED 168 336)
************
337
```

```
(TEST-COMPLETED 12 337)
************
338
(TEST-COMPLETED 169 338)
************
339
(TEST-COMPLETED 18 339)
************
340
(TEST-COMPLETED 170 340)
************
341
(TEST-COMPLETED 24 341)
************
342
(TEST-COMPLETED 171 342)
************
343
(TEST-COMPLETED 30 343)
************
344
(TEST-COMPLETED 172 344)
************
345
(TEST-COMPLETED 36 345)
************
346
(TEST-COMPLETED 173 346)
************
347
(TEST-COMPLETED 42 347)
************
348
(TEST-COMPLETED 174 348)
************
349
(TEST-COMPLETED 48 349)
************
350
(TEST-COMPLETED 175 350)
************
351
(TEST-COMPLETED 54 351)
************
352
(TEST-COMPLETED 176 352)
************
353
(TEST-COMPLETED 60 353)
************
354
(TEST-COMPLETED 177 354)
************
355
(TEST-COMPLETED 66 355)
************
356
```

```
(TEST-COMPLETED 178 356)
************
357
(TEST-COMPLETED 72 357)
************
358
(TEST-COMPLETED 179 358)
************
359
(TEST-COMPLETED 78 359)
************
360
(TEST-COMPLETED 180 360)
************
361
(TEST-COMPLETED 84 361)
************
362
(TEST-COMPLETED 181 362)
************
363
(TEST-COMPLETED 90 363)
************
364
(TEST-COMPLETED 182 364)
************
365
(TEST-COMPLETED 96 365)
************
366
(TEST-COMPLETED 183 366)
************
367
(TEST-COMPLETED 102 367)
************
368
(TEST-COMPLETED 184 368)
************
369
(TEST-COMPLETED 108 369)
************
370
(TEST-COMPLETED 185 370)
************
371
(TEST-COMPLETED 114 371)
************
372
(TEST-COMPLETED 186 372)
************
373
(TEST-COMPLETED 120 373)
************
374
(TEST-COMPLETED 187 374)
************
375
```

```
(TEST-COMPLETED 126 375)
************
376
(TEST-COMPLETED 188 376)
************
377
(TEST-COMPLETED 132 377)
************
378
(TEST-COMPLETED 189 378)
************
379
(TEST-COMPLETED 138 379)
************
380
(TEST-COMPLETED 190 380)
************
381
(TEST-COMPLETED 144 381)
************
382
(TEST-COMPLETED 191 382)
************
383
(TEST-COMPLETED 150 383)
************
384
(TEST-COMPLETED 192 384)
************
385
(TEST-COMPLETED 156 385)
************
386
(TEST-COMPLETED 193 386)
************
387
(TEST-COMPLETED 162 387)
************
388
(TEST-COMPLETED 194 388)
************
389
(TEST-COMPLETED 168 389)
************
390
(TEST-COMPLETED 195 390)
************
391
(TEST-COMPLETED 174 391)
************
392
(TEST-COMPLETED 196 392)
************
393
(TEST-COMPLETED 180 393)
************
394
```

```
(TEST-COMPLETED 197 394)
************
395
(TEST-COMPLETED 186 395)
************
396
(TEST-COMPLETED 198 396)
************
397
(TEST-COMPLETED 192 397)
************
398
(TEST-COMPLETED 199 398)
************
399
(TEST-COMPLETED 198 399)
************
400
(TEST-COMPLETED 200 400)
************
401
(TEST-COMPLETED 204 401)
************
402
(TEST-COMPLETED 201 402)
************
403
(TEST-COMPLETED 210 403)
************
404
(TEST-COMPLETED 202 404)
************
405
(TEST-COMPLETED 216 405)
************
406
(TEST-COMPLETED 203 406)
************
407
(TEST-COMPLETED 222 407)
************
408
(TEST-COMPLETED 204 408)
************
409
(TEST-COMPLETED 228 409)
************
410
(TEST-COMPLETED 205 410)
************
411
(TEST-COMPLETED 234 411)
************
412
(TEST-COMPLETED 206 412)
************
413
```

```
(TEST-COMPLETED 240 413)
************
414
(TEST-COMPLETED 207 414)
************
415
(TEST-COMPLETED 246 415)
************
416
(TEST-COMPLETED 208 416)
************
417
(TEST-COMPLETED 252 417)
************
418
(TEST-COMPLETED 209 418)
************
419
(TEST-COMPLETED 258 419)
************
420
(TEST-COMPLETED 210 420)
************
421
(TEST-COMPLETED 264 421)
************
422
(TEST-COMPLETED 211 422)
************
423
(TEST-COMPLETED 270 423)
************
424
(TEST-COMPLETED 212 424)
************
425
(TEST-COMPLETED 276 425)
************
426
(TEST-COMPLETED 213 426)
************
427
(TEST-COMPLETED 282 427)
************
428
(TEST-COMPLETED 214 428)
************
429
(TEST-COMPLETED 288 429)
************
430
(TEST-COMPLETED 215 430)
************
431
(TEST-COMPLETED 294 431)
************
432
```

```
(TEST-COMPLETED 216 432)
************
433
(TEST-COMPLETED 300 433)
************
434
(TEST-COMPLETED 217 434)
************
435
(TEST-COMPLETED 306 435)
************
436
(TEST-COMPLETED 218 436)
************
437
(TEST-COMPLETED 312 437)
************
438
(TEST-COMPLETED 219 438)
************
439
(TEST-COMPLETED 318 439)
************
440
(TEST-COMPLETED 220 440)
************
441
(TEST-COMPLETED 324 441)
************
442
(TEST-COMPLETED 221 442)
************
443
(TEST-COMPLETED 330 443)
************
444
(TEST-COMPLETED 222 444)
************
445
(TEST-COMPLETED 336 445)
************
446
(TEST-COMPLETED 223 446)
************
447
(TEST-COMPLETED 342 447)
************
448
(TEST-COMPLETED 224 448)
************
449
(TEST-COMPLETED 348 449)
************
450
(TEST-COMPLETED 225 450)
************
451
```

```
(TEST-COMPLETED 354 451)
************
452
(TEST-COMPLETED 226 452)
************
453
(TEST-COMPLETED 360 453)
************
454
(TEST-COMPLETED 227 454)
************
455
(TEST-COMPLETED 366 455)
************
456
(TEST-COMPLETED 228 456)
************
457
(TEST-COMPLETED 372 457)
************
458
(TEST-COMPLETED 229 458)
************
459
(TEST-COMPLETED 378 459)
************
460
(TEST-COMPLETED 230 460)
************
461
(TEST-COMPLETED 384 461)
************
462
(TEST-COMPLETED 231 462)
************
463
(TEST-COMPLETED 390 463)
************
464
(TEST-COMPLETED 232 464)
************
465
(TEST-COMPLETED 396 465)
************
466
(TEST-COMPLETED 233 466)
************
467
(TEST-COMPLETED 402 467)
************
468
(TEST-COMPLETED 234 468)
************
469
(TEST-COMPLETED 408 469)
************
470
```

```
(TEST-COMPLETED 235 470)
************
471
(TEST-COMPLETED 414 471)
************
472
(TEST-COMPLETED 236 472)
************
473
(TEST-COMPLETED 420 473)
************
474
(TEST-COMPLETED 237 474)
************
475
(TEST-COMPLETED 426 475)
************
476
(TEST-COMPLETED 238 476)
************
477
(EVEN 432 477)
(BRANCHING 216)
(TEST-COMPLETED 108 216)
(BRANCHING 716)
(TEST-COMPLETED 358 716)
************
478
(TEST-COMPLETED 239 478)
************
479
(EVEN 438 479)
(BRANCHING 219)
(EVEN 658 219)
(BRANCHING 329)
(EVEN 988 329)
(BRANCHING 494)
(TEST-COMPLETED 247 494)
(BRANCHING 994)
(TEST-COMPLETED 497 994)
(BRANCHING 829)
(TEST-COMPLETED 488 829)
(BRANCHING 719)
(TEST-COMPLETED 158 719)
************
480
(TEST-COMPLETED 240 480)
************
481
(EVEN 444 481)
(BRANCHING 222)
(TEST-COMPLETED 111 222)
(BRANCHING 722)
(TEST-COMPLETED 361 722)
************
482
(TEST-COMPLETED 241 482)
```

```
***********
483
(EVEN 450 483)
(BRANCHING 225)
(EVEN 676 225)
(BRANCHING 338)
(TEST-COMPLETED 169 338)
(BRANCHING 838)
(TEST-COMPLETED 419 838)
(BRANCHING 725)
(TEST-COMPLETED 176 725)
***********
484
(TEST-COMPLETED 242 484)
***********
485
(EVEN 456 485)
(BRANCHING 228)
(TEST-COMPLETED 114 228)
(BRANCHING 728)
(TEST-COMPLETED 364 728)
***********
486
(TEST-COMPLETED 243 486)
***********
487
(EVEN 462 487)
(BRANCHING 231)
(EVEN 694 231)
(BRANCHING 347)
(TEST-COMPLETED 42 347)
(BRANCHING 847)
(TEST-COMPLETED 542 847)
(BRANCHING 731)
(TEST-COMPLETED 194 731)
***********
488
(TEST-COMPLETED 244 488)
***********
489
(EVEN 468 489)
(BRANCHING 234)
(TEST-COMPLETED 117 234)
(BRANCHING 734)
(TEST-COMPLETED 367 734)
***********
490
(TEST-COMPLETED 245 490)
***********
491
(EVEN 474 491)
(BRANCHING 237)
(EVEN 712 237)
(BRANCHING 356)
(TEST-COMPLETED 178 356)
(BRANCHING 856)
(TEST-COMPLETED 428 856)
```

```
(BRANCHING 737)
(TEST-COMPLETED 212 737)
* * * * * * * * * * * *
492
(TEST-COMPLETED 246 492)
* * * * * * * * * * * *
493
(EVEN 480 493)
(BRANCHING 240)
(TEST-COMPLETED 120 240)
(BRANCHING 740)
(TEST-COMPLETED 370 740)
* * * * * * * * * * * *
494
(TEST-COMPLETED 247 494)
* * * * * * * * * * * *
495
(EVEN 486 495)
(BRANCHING 243)
(EVEN 730 243)
(BRANCHING 365)
(TEST-COMPLETED 96 365)
(BRANCHING 865)
(TEST-COMPLETED 596 865)
(BRANCHING 743)
(TEST-COMPLETED 230 743)
* * * * * * * * * * * *
496
(TEST-COMPLETED 248 496)
* * * * * * * * * * * *
497
(EVEN 492 497)
(BRANCHING 246)
(TEST-COMPLETED 123 246)
(BRANCHING 746)
(TEST-COMPLETED 373 746)
* * * * * * * * * * * *
498
(TEST-COMPLETED 249 498)
* * * * * * * * * * * *
499
(EVEN 498 499)
(BRANCHING 249)
(EVEN 748 249)
(BRANCHING 374)
(TEST-COMPLETED 187 374)
(BRANCHING 874)
(TEST-COMPLETED 437 874)
(BRANCHING 749)
(TEST-COMPLETED 248 749)
* * * * * * * * * * * *
500
(TEST-COMPLETED 250 500)
* * * * * * * * * * * *
501
(EVEN 504 501)
(BRANCHING 252)
```

```
(TEST-COMPLETED 126 252)
(BRANCHING 752)
(TEST-COMPLETED 376 752)
************
502
(TEST-COMPLETED 251 502)
************
503
(EVEN 510 503)
(BRANCHING 255)
(EVEN 766 255)
(BRANCHING 383)
(TEST-COMPLETED 150 383)
(BRANCHING 883)
(TEST-COMPLETED 650 883)
(BRANCHING 755)
(TEST-COMPLETED 266 755)
************
504
(TEST-COMPLETED 252 504)
************
505
(EVEN 516 505)
(BRANCHING 258)
(TEST-COMPLETED 129 258)
(BRANCHING 758)
(TEST-COMPLETED 379 758)
************
506
(TEST-COMPLETED 253 506)
************
507
(EVEN 522 507)
(BRANCHING 261)
(EVEN 784 261)
(BRANCHING 392)
(TEST-COMPLETED 196 392)
(BRANCHING 892)
(TEST-COMPLETED 446 892)
(BRANCHING 761)
(TEST-COMPLETED 284 761)
************
508
(TEST-COMPLETED 254 508)
************
509
(EVEN 528 509)
(BRANCHING 264)
(TEST-COMPLETED 132 264)
(BRANCHING 764)
(TEST-COMPLETED 382 764)
************
510
(TEST-COMPLETED 255 510)
************
511
(EVEN 534 511)
```

```
(BRANCHING 267)
(EVEN 802 267)
(BRANCHING 401)
(TEST-COMPLETED 204 401)
(BRANCHING 901)
(TEST-COMPLETED 704 901)
(BRANCHING 767)
(TEST-COMPLETED 302 767)
* * * * * * * * * * * *
512
(TEST-COMPLETED 256 512)
* * * * * * * * * * * *
513
(EVEN 540 513)
(BRANCHING 270)
(TEST-COMPLETED 135 270)
(BRANCHING 770)
(TEST-COMPLETED 385 770)
* * * * * * * * * * * *
514
(TEST-COMPLETED 257 514)
* * * * * * * * * * * *
515
(EVEN 546 515)
(BRANCHING 273)
(EVEN 820 273)
(BRANCHING 410)
(TEST-COMPLETED 205 410)
(BRANCHING 910)
(TEST-COMPLETED 455 910)
(BRANCHING 773)
(TEST-COMPLETED 320 773)
* * * * * * * * * * * *
516
(TEST-COMPLETED 258 516)
* * * * * * * * * * * *
517
(EVEN 552 517)
(BRANCHING 276)
(TEST-COMPLETED 138 276)
(BRANCHING 776)
(TEST-COMPLETED 388 776)
* * * * * * * * * * * *
518
(TEST-COMPLETED 259 518)
* * * * * * * * * * * *
519
(EVEN 558 519)
(BRANCHING 279)
(EVEN 838 279)
(BRANCHING 419)
(TEST-COMPLETED 258 419)
(BRANCHING 919)
(TEST-COMPLETED 758 919)
(BRANCHING 779)
(TEST-COMPLETED 338 779)
* * * * * * * * * * * *
```

```
520
(TEST-COMPLETED 260 520)
************
521
(EVEN 564 521)
(BRANCHING 282)
(TEST-COMPLETED 141 282)
(BRANCHING 782)
(TEST-COMPLETED 391 782)
************
522
(TEST-COMPLETED 261 522)
************
523
(EVEN 570 523)
(BRANCHING 285)
(EVEN 856 285)
(BRANCHING 428)
(TEST-COMPLETED 214 428)
(BRANCHING 928)
(TEST-COMPLETED 464 928)
(BRANCHING 785)
(TEST-COMPLETED 356 785)
************
524
(TEST-COMPLETED 262 524)
************
525
(EVEN 576 525)
(BRANCHING 288)
(TEST-COMPLETED 144 288)
(BRANCHING 788)
(TEST-COMPLETED 394 788)
************
526
(TEST-COMPLETED 263 526)
************
527
(EVEN 582 527)
(BRANCHING 291)
(EVEN 874 291)
(BRANCHING 437)
(TEST-COMPLETED 312 437)
(BRANCHING 937)
(TEST-COMPLETED 812 937)
(BRANCHING 791)
(TEST-COMPLETED 374 791)
************
528
(TEST-COMPLETED 264 528)
************
529
(EVEN 588 529)
(BRANCHING 294)
(TEST-COMPLETED 147 294)
(BRANCHING 794)
(TEST-COMPLETED 397 794)
```

```
************
530
(TEST-COMPLETED 265 530)
************
531
(EVEN 594 531)
(BRANCHING 297)
(EVEN 892 297)
(BRANCHING 446)
(TEST-COMPLETED 223 446)
(BRANCHING 946)
(TEST-COMPLETED 473 946)
(BRANCHING 797)
(TEST-COMPLETED 392 797)
************
532
(TEST-COMPLETED 266 532)
************
533
(EVEN 600 533)
(BRANCHING 300)
(TEST-COMPLETED 150 300)
(BRANCHING 800)
(TEST-COMPLETED 400 800)
************
534
(TEST-COMPLETED 267 534)
************
535
(EVEN 606 535)
(BRANCHING 303)
(EVEN 910 303)
(BRANCHING 455)
(TEST-COMPLETED 366 455)
(BRANCHING 955)
(EVEN 866 955)
(BRANCHING 433)
(TEST-COMPLETED 300 433)
(BRANCHING 933)
(TEST-COMPLETED 800 933)
(BRANCHING 803)
(TEST-COMPLETED 410 803)
************
536
(TEST-COMPLETED 268 536)
************
537
(EVEN 612 537)
(BRANCHING 306)
(TEST-COMPLETED 153 306)
(BRANCHING 806)
(TEST-COMPLETED 403 806)
************
538
(TEST-COMPLETED 269 538)
************
539
```

```
(EVEN 618 539)
(BRANCHING 309)
(EVEN 928 309)
(BRANCHING 464)
(TEST-COMPLETED 232 464)
(BRANCHING 964)
(TEST-COMPLETED 482 964)
(BRANCHING 809)
(TEST-COMPLETED 428 809)
************
540
(TEST-COMPLETED 270 540)
************
541
(EVEN 624 541)
(BRANCHING 312)
(TEST-COMPLETED 156 312)
(BRANCHING 812)
(TEST-COMPLETED 406 812)
************
542
(TEST-COMPLETED 271 542)
************
543
(EVEN 630 543)
(BRANCHING 315)
(EVEN 946 315)
(BRANCHING 473)
(TEST-COMPLETED 420 473)
(BRANCHING 973)
(EVEN 920 973)
(BRANCHING 460)
(TEST-COMPLETED 230 460)
(BRANCHING 960)
(TEST-COMPLETED 480 960)
(BRANCHING 815)
(TEST-COMPLETED 446 815)
************
544
(TEST-COMPLETED 272 544)
************
545
(EVEN 636 545)
(BRANCHING 318)
(TEST-COMPLETED 159 318)
(BRANCHING 818)
(TEST-COMPLETED 409 818)
************
546
(TEST-COMPLETED 273 546)
************
547
(EVEN 642 547)
(BRANCHING 321)
(EVEN 964 321)
(BRANCHING 482)
(TEST-COMPLETED 241 482)
```

```
(BRANCHING 982)
(TEST-COMPLETED 491 982)
(BRANCHING 821)
(TEST-COMPLETED 464 821)
************
548
(TEST-COMPLETED 274 548)
************
549
(EVEN 648 549)
(BRANCHING 324)
(TEST-COMPLETED 162 324)
(BRANCHING 824)
(TEST-COMPLETED 412 824)
************
550
(TEST-COMPLETED 275 550)
************
551
(EVEN 654 551)
(BRANCHING 327)
(EVEN 982 327)
(BRANCHING 491)
(EVEN 474 491)
(BRANCHING 237)
(EVEN 712 237)
(BRANCHING 356)
(TEST-COMPLETED 178 356)
(BRANCHING 856)
(TEST-COMPLETED 428 856)
(BRANCHING 737)
(TEST-COMPLETED 212 737)
(BRANCHING 991)
(EVEN 974 991)
(BRANCHING 487)
(EVEN 462 487)
(BRANCHING 231)
(EVEN 694 231)
(BRANCHING 347)
(TEST-COMPLETED 42 347)
(BRANCHING 847)
(TEST-COMPLETED 542 847)
(BRANCHING 731)
(TEST-COMPLETED 194 731)
(BRANCHING 987)
(EVEN 962 987)
(BRANCHING 481)
(EVEN 444 481)
(BRANCHING 222)
(TEST-COMPLETED 111 222)
(BRANCHING 722)
(TEST-COMPLETED 361 722)
(BRANCHING 981)
(EVEN 944 981)
(BRANCHING 472)
(TEST-COMPLETED 236 472)
(BRANCHING 972)
```

```
(TEST-COMPLETED 486 972)
(BRANCHING 827)
(TEST-COMPLETED 482 827)
* * * * * * * * * * * *
552
(TEST-COMPLETED 276 552)
* * * * * * * * * * * *
553
(EVEN 660 553)
(BRANCHING 330)
(TEST-COMPLETED 165 330)
(BRANCHING 830)
(TEST-COMPLETED 415 830)
* * * * * * * * * * * *
554
(TEST-COMPLETED 277 554)
* * * * * * * * * * * *
555
(EVEN 666 555)
(BRANCHING 333)
(TEST-COMPLETED 0 333)
(BRANCHING 833)
(TEST-COMPLETED 500 833)
* * * * * * * * * * * *
556
(TEST-COMPLETED 278 556)
* * * * * * * * * * * *
557
(EVEN 672 557)
(BRANCHING 336)
(TEST-COMPLETED 168 336)
(BRANCHING 836)
(TEST-COMPLETED 418 836)
* * * * * * * * * * * *
558
(TEST-COMPLETED 279 558)
* * * * * * * * * * * *
559
(EVEN 678 559)
(BRANCHING 339)
(TEST-COMPLETED 18 339)
(BRANCHING 839)
(TEST-COMPLETED 518 839)
* * * * * * * * * * * *
560
(TEST-COMPLETED 280 560)
* * * * * * * * * * * *
561
(EVEN 684 561)
(BRANCHING 342)
(TEST-COMPLETED 171 342)
(BRANCHING 842)
(TEST-COMPLETED 421 842)
* * * * * * * * * * * *
562
(TEST-COMPLETED 281 562)
* * * * * * * * * * * *
```

```
563
(EVEN 690 563)
(BRANCHING 345)
(TEST-COMPLETED 36 345)
(BRANCHING 845)
(TEST-COMPLETED 536 845)
************
564
(TEST-COMPLETED 282 564)
************
565
(EVEN 696 565)
(BRANCHING 348)
(TEST-COMPLETED 174 348)
(BRANCHING 848)
(TEST-COMPLETED 424 848)
************
566
(TEST-COMPLETED 283 566)
************
567
(EVEN 702 567)
(BRANCHING 351)
(TEST-COMPLETED 54 351)
(BRANCHING 851)
(TEST-COMPLETED 554 851)
************
568
(TEST-COMPLETED 284 568)
************
569
(EVEN 708 569)
(BRANCHING 354)
(TEST-COMPLETED 177 354)
(BRANCHING 854)
(TEST-COMPLETED 427 854)
************
570
(TEST-COMPLETED 285 570)
************
571
(EVEN 714 571)
(BRANCHING 357)
(TEST-COMPLETED 72 357)
(BRANCHING 857)
(TEST-COMPLETED 572 857)
************
572
(TEST-COMPLETED 286 572)
************
573
(EVEN 720 573)
(BRANCHING 360)
(TEST-COMPLETED 180 360)
(BRANCHING 860)
(TEST-COMPLETED 430 860)
************
```

```
574
(TEST-COMPLETED 287 574)
************
575
(EVEN 726 575)
(BRANCHING 363)
(TEST-COMPLETED 90 363)
(BRANCHING 863)
(TEST-COMPLETED 590 863)
************
576
(TEST-COMPLETED 288 576)
************
577
(EVEN 732 577)
(BRANCHING 366)
(TEST-COMPLETED 183 366)
(BRANCHING 866)
(TEST-COMPLETED 433 866)
************
578
(TEST-COMPLETED 289 578)
************
579
(EVEN 738 579)
(BRANCHING 369)
(TEST-COMPLETED 108 369)
(BRANCHING 869)
(TEST-COMPLETED 608 869)
************
580
(TEST-COMPLETED 290 580)
************
581
(EVEN 744 581)
(BRANCHING 372)
(TEST-COMPLETED 186 372)
(BRANCHING 872)
(TEST-COMPLETED 436 872)
************
582
(TEST-COMPLETED 291 582)
************
583
(EVEN 750 583)
(BRANCHING 375)
(TEST-COMPLETED 126 375)
(BRANCHING 875)
(TEST-COMPLETED 626 875)
************
584
(TEST-COMPLETED 292 584)
************
585
(EVEN 756 585)
(BRANCHING 378)
(TEST-COMPLETED 189 378)
```

```
(BRANCHING 878)
(TEST-COMPLETED 439 878)
* * * * * * * * * * * *
586
(TEST-COMPLETED 293 586)
* * * * * * * * * * * *
587
(EVEN 762 587)
(BRANCHING 381)
(TEST-COMPLETED 144 381)
(BRANCHING 881)
(TEST-COMPLETED 644 881)
* * * * * * * * * * * *
588
(TEST-COMPLETED 294 588)
* * * * * * * * * * * *
589
(EVEN 768 589)
(BRANCHING 384)
(TEST-COMPLETED 192 384)
(BRANCHING 884)
(TEST-COMPLETED 442 884)
* * * * * * * * * * * *
590
(TEST-COMPLETED 295 590)
* * * * * * * * * * * *
591
(EVEN 774 591)
(BRANCHING 387)
(TEST-COMPLETED 162 387)
(BRANCHING 887)
(TEST-COMPLETED 662 887)
* * * * * * * * * * * *
592
(TEST-COMPLETED 296 592)
* * * * * * * * * * * *
593
(EVEN 780 593)
(BRANCHING 390)
(TEST-COMPLETED 195 390)
(BRANCHING 890)
(TEST-COMPLETED 445 890)
* * * * * * * * * * * *
594
(TEST-COMPLETED 297 594)
* * * * * * * * * * * *
595
(EVEN 786 595)
(BRANCHING 393)
(TEST-COMPLETED 180 393)
(BRANCHING 893)
(TEST-COMPLETED 680 893)
* * * * * * * * * * * *
596
(TEST-COMPLETED 298 596)
* * * * * * * * * * * *
597
```

```
(EVEN 792 597)
(BRANCHING 396)
(TEST-COMPLETED 198 396)
(BRANCHING 896)
(TEST-COMPLETED 448 896)
************
598
(TEST-COMPLETED 299 598)
************
599
(EVEN 798 599)
(BRANCHING 399)
(TEST-COMPLETED 198 399)
(BRANCHING 899)
(TEST-COMPLETED 698 899)
************
600
(TEST-COMPLETED 300 600)
************
601
(EVEN 804 601)
(BRANCHING 402)
(TEST-COMPLETED 201 402)
(BRANCHING 902)
(TEST-COMPLETED 451 902)
************
602
(TEST-COMPLETED 301 602)
************
603
(EVEN 810 603)
(BRANCHING 405)
(TEST-COMPLETED 216 405)
(BRANCHING 905)
(TEST-COMPLETED 716 905)
************
604
(TEST-COMPLETED 302 604)
************
605
(EVEN 816 605)
(BRANCHING 408)
(TEST-COMPLETED 204 408)
(BRANCHING 908)
(TEST-COMPLETED 454 908)
************
606
(TEST-COMPLETED 303 606)
************
607
(EVEN 822 607)
(BRANCHING 411)
(TEST-COMPLETED 234 411)
(BRANCHING 911)
(TEST-COMPLETED 734 911)
************
608
```

```
(TEST-COMPLETED 304 608)
************
609
(EVEN 828 609)
(BRANCHING 414)
(TEST-COMPLETED 207 414)
(BRANCHING 914)
(TEST-COMPLETED 457 914)
************
610
(TEST-COMPLETED 305 610)
************
611
(EVEN 834 611)
(BRANCHING 417)
(TEST-COMPLETED 252 417)
(BRANCHING 917)
(TEST-COMPLETED 752 917)
************
612
(TEST-COMPLETED 306 612)
************
613
(EVEN 840 613)
(BRANCHING 420)
(TEST-COMPLETED 210 420)
(BRANCHING 920)
(TEST-COMPLETED 460 920)
************
614
(TEST-COMPLETED 307 614)
************
615
(EVEN 846 615)
(BRANCHING 423)
(TEST-COMPLETED 270 423)
(BRANCHING 923)
(TEST-COMPLETED 770 923)
************
616
(TEST-COMPLETED 308 616)
************
617
(EVEN 852 617)
(BRANCHING 426)
(TEST-COMPLETED 213 426)
(BRANCHING 926)
(TEST-COMPLETED 463 926)
************
618
(TEST-COMPLETED 309 618)
************
619
(EVEN 858 619)
(BRANCHING 429)
(TEST-COMPLETED 288 429)
(BRANCHING 929)
```

```
(TEST-COMPLETED 788 929)
************
620
(TEST-COMPLETED 310 620)
************
621
(EVEN 864 621)
(BRANCHING 432)
(TEST-COMPLETED 216 432)
(BRANCHING 932)
(TEST-COMPLETED 466 932)
************
622
(TEST-COMPLETED 311 622)
************
623
(EVEN 870 623)
(BRANCHING 435)
(TEST-COMPLETED 306 435)
(BRANCHING 935)
(TEST-COMPLETED 806 935)
************
624
(TEST-COMPLETED 312 624)
************
625
(EVEN 876 625)
(BRANCHING 438)
(TEST-COMPLETED 219 438)
(BRANCHING 938)
(TEST-COMPLETED 469 938)
************
626
(TEST-COMPLETED 313 626)
************
627
(EVEN 882 627)
(BRANCHING 441)
(TEST-COMPLETED 324 441)
(BRANCHING 941)
(TEST-COMPLETED 824 941)
************
628
(TEST-COMPLETED 314 628)
************
629
(EVEN 888 629)
(BRANCHING 444)
(TEST-COMPLETED 222 444)
(BRANCHING 944)
(TEST-COMPLETED 472 944)
************
630
(TEST-COMPLETED 315 630)
************
631
(EVEN 894 631)
```

```
(BRANCHING 447)
(TEST-COMPLETED 342 447)
(BRANCHING 947)
(TEST-COMPLETED 842 947)
************
632
(TEST-COMPLETED 316 632)
************
633
(EVEN 900 633)
(BRANCHING 450)
(TEST-COMPLETED 225 450)
(BRANCHING 950)
(TEST-COMPLETED 475 950)
************
634
(TEST-COMPLETED 317 634)
************
635
(EVEN 906 635)
(BRANCHING 453)
(TEST-COMPLETED 360 453)
(BRANCHING 953)
(EVEN 860 953)
(BRANCHING 430)
(TEST-COMPLETED 215 430)
(BRANCHING 930)
(TEST-COMPLETED 465 930)
************
636
(TEST-COMPLETED 318 636)
************
637
(EVEN 912 637)
(BRANCHING 456)
(TEST-COMPLETED 228 456)
(BRANCHING 956)
(TEST-COMPLETED 478 956)
************
638
(TEST-COMPLETED 319 638)
************
639
(EVEN 918 639)
(BRANCHING 459)
(TEST-COMPLETED 378 459)
(BRANCHING 959)
(EVEN 878 959)
(BRANCHING 439)
(TEST-COMPLETED 318 439)
(BRANCHING 939)
(TEST-COMPLETED 818 939)
************
640
(TEST-COMPLETED 320 640)
************
641
```

```
(EVEN 924 641)
(BRANCHING 462)
(TEST-COMPLETED 231 462)
(BRANCHING 962)
(TEST-COMPLETED 481 962)
* * * * * * * * * * * *
642
(TEST-COMPLETED 321 642)
* * * * * * * * * * * *
643
(EVEN 930 643)
(BRANCHING 465)
(TEST-COMPLETED 396 465)
(BRANCHING 965)
(EVEN 896 965)
(BRANCHING 448)
(TEST-COMPLETED 224 448)
(BRANCHING 948)
(TEST-COMPLETED 474 948)
* * * * * * * * * * * *
644
(TEST-COMPLETED 322 644)
* * * * * * * * * * * *
645
(EVEN 936 645)
(BRANCHING 468)
(TEST-COMPLETED 234 468)
(BRANCHING 968)
(TEST-COMPLETED 484 968)
* * * * * * * * * * * *
646
(TEST-COMPLETED 323 646)
* * * * * * * * * * * *
647
(EVEN 942 647)
(BRANCHING 471)
(TEST-COMPLETED 414 471)
(BRANCHING 971)
(EVEN 914 971)
(BRANCHING 457)
(TEST-COMPLETED 372 457)
(BRANCHING 957)
(EVEN 872 957)
(BRANCHING 436)
(TEST-COMPLETED 218 436)
(BRANCHING 936)
(TEST-COMPLETED 468 936)
* * * * * * * * * * * *
648
(TEST-COMPLETED 324 648)
* * * * * * * * * * * *
649
(EVEN 948 649)
(BRANCHING 474)
(TEST-COMPLETED 237 474)
(BRANCHING 974)
(TEST-COMPLETED 487 974)
```

```
************
650
(TEST-COMPLETED 325 650)
************
651
(EVEN 954 651)
(BRANCHING 477)
(EVEN 432 477)
(BRANCHING 216)
(TEST-COMPLETED 108 216)
(BRANCHING 716)
(TEST-COMPLETED 358 716)
(BRANCHING 977)
(EVEN 932 977)
(BRANCHING 466)
(TEST-COMPLETED 233 466)
(BRANCHING 966)
(TEST-COMPLETED 483 966)
************
652
(TEST-COMPLETED 326 652)
************
653
(EVEN 960 653)
(BRANCHING 480)
(TEST-COMPLETED 240 480)
(BRANCHING 980)
(TEST-COMPLETED 490 980)
************
654
(TEST-COMPLETED 327 654)
************
655
(EVEN 966 655)
(BRANCHING 483)
(EVEN 450 483)
(BRANCHING 225)
(EVEN 676 225)
(BRANCHING 338)
(TEST-COMPLETED 169 338)
(BRANCHING 838)
(TEST-COMPLETED 419 838)
(BRANCHING 725)
(TEST-COMPLETED 176 725)
(BRANCHING 983)
(EVEN 950 983)
(BRANCHING 475)
(TEST-COMPLETED 426 475)
(BRANCHING 975)
(EVEN 926 975)
(BRANCHING 463)
(TEST-COMPLETED 390 463)
(BRANCHING 963)
(EVEN 890 963)
(BRANCHING 445)
(TEST-COMPLETED 336 445)
(BRANCHING 945)
```

```
(TEST-COMPLETED 836 945)
************
656
(TEST-COMPLETED 328 656)
************
657
(EVEN 972 657)
(BRANCHING 486)
(TEST-COMPLETED 243 486)
(BRANCHING 986)
(TEST-COMPLETED 493 986)
************
658
(TEST-COMPLETED 329 658)
************
659
(EVEN 978 659)
(BRANCHING 489)
(EVEN 468 489)
(BRANCHING 234)
(TEST-COMPLETED 117 234)
(BRANCHING 734)
(TEST-COMPLETED 367 734)
(BRANCHING 989)
(EVEN 968 989)
(BRANCHING 484)
(TEST-COMPLETED 242 484)
(BRANCHING 984)
(TEST-COMPLETED 492 984)
************
660
(TEST-COMPLETED 330 660)
************
661
(EVEN 984 661)
(BRANCHING 492)
(TEST-COMPLETED 246 492)
(BRANCHING 992)
(TEST-COMPLETED 496 992)
************
662
(TEST-COMPLETED 331 662)
************
663
(EVEN 990 663)
(BRANCHING 495)
(EVEN 486 495)
(BRANCHING 243)
(EVEN 730 243)
(BRANCHING 365)
(TEST-COMPLETED 96 365)
(BRANCHING 865)
(TEST-COMPLETED 596 865)
(BRANCHING 743)
(TEST-COMPLETED 230 743)
(BRANCHING 995)
(EVEN 986 995)
```

```
(BRANCHING 493)
(EVEN 480 493)
(BRANCHING 240)
(TEST-COMPLETED 120 240)
(BRANCHING 740)
(TEST-COMPLETED 370 740)
(BRANCHING 993)
(EVEN 980 993)
(BRANCHING 490)
(TEST-COMPLETED 245 490)
(BRANCHING 990)
(TEST-COMPLETED 495 990)
************
664
(TEST-COMPLETED 332 664)
************
665
(EVEN 996 665)
(BRANCHING 498)
(TEST-COMPLETED 249 498)
(BRANCHING 998)
(TEST-COMPLETED 499 998)
************
666
(TEST-COMPLETED 333 666)
************
667
(TEST-COMPLETED 2 667)
************
668
(TEST-COMPLETED 334 668)
************
669
(TEST-COMPLETED 8 669)
************
670
(TEST-COMPLETED 335 670)
************
671
(TEST-COMPLETED 14 671)
************
672
(TEST-COMPLETED 336 672)
************
673
(TEST-COMPLETED 20 673)
************
674
(TEST-COMPLETED 337 674)
************
675
(TEST-COMPLETED 26 675)
************
676
(TEST-COMPLETED 338 676)
************
677
```

```
(TEST-COMPLETED 32 677)
************
678
(TEST-COMPLETED 339 678)
************
679
(TEST-COMPLETED 38 679)
************
680
(TEST-COMPLETED 340 680)
************
681
(TEST-COMPLETED 44 681)
************
682
(TEST-COMPLETED 341 682)
************
683
(TEST-COMPLETED 50 683)
************
684
(TEST-COMPLETED 342 684)
************
685
(TEST-COMPLETED 56 685)
************
686
(TEST-COMPLETED 343 686)
************
687
(TEST-COMPLETED 62 687)
************
688
(TEST-COMPLETED 344 688)
************
689
(TEST-COMPLETED 68 689)
************
690
(TEST-COMPLETED 345 690)
************
691
(TEST-COMPLETED 74 691)
************
692
(TEST-COMPLETED 346 692)
************
693
(TEST-COMPLETED 80 693)
************
694
(TEST-COMPLETED 347 694)
************
695
(TEST-COMPLETED 86 695)
************
696
```

```
(TEST-COMPLETED 348 696)
************
697
(TEST-COMPLETED 92 697)
************
698
(TEST-COMPLETED 349 698)
************
699
(TEST-COMPLETED 98 699)
************
700
(TEST-COMPLETED 350 700)
************
701
(TEST-COMPLETED 104 701)
************
702
(TEST-COMPLETED 351 702)
************
703
(TEST-COMPLETED 110 703)
************
704
(TEST-COMPLETED 352 704)
************
705
(TEST-COMPLETED 116 705)
************
706
(TEST-COMPLETED 353 706)
************
707
(TEST-COMPLETED 122 707)
************
708
(TEST-COMPLETED 354 708)
************
709
(TEST-COMPLETED 128 709)
************
710
(TEST-COMPLETED 355 710)
************
711
(TEST-COMPLETED 134 711)
************
712
(TEST-COMPLETED 356 712)
************
713
(TEST-COMPLETED 140 713)
************
714
(TEST-COMPLETED 357 714)
************
715
```

```
(TEST-COMPLETED 146 715)
************
716
(TEST-COMPLETED 358 716)
************
717
(TEST-COMPLETED 152 717)
************
718
(TEST-COMPLETED 359 718)
************
719
(TEST-COMPLETED 158 719)
************
720
(TEST-COMPLETED 360 720)
************
721
(TEST-COMPLETED 164 721)
************
722
(TEST-COMPLETED 361 722)
************
723
(TEST-COMPLETED 170 723)
************
724
(TEST-COMPLETED 362 724)
************
725
(TEST-COMPLETED 176 725)
************
726
(TEST-COMPLETED 363 726)
************
727
(TEST-COMPLETED 182 727)
************
728
(TEST-COMPLETED 364 728)
************
729
(TEST-COMPLETED 188 729)
************
730
(TEST-COMPLETED 365 730)
************
731
(TEST-COMPLETED 194 731)
************
732
(TEST-COMPLETED 366 732)
************
733
(TEST-COMPLETED 200 733)
************
734
```

```
(TEST-COMPLETED 367 734)
************
735
(TEST-COMPLETED 206 735)
************
736
(TEST-COMPLETED 368 736)
************
737
(TEST-COMPLETED 212 737)
************
738
(TEST-COMPLETED 369 738)
************
739
(TEST-COMPLETED 218 739)
************
740
(TEST-COMPLETED 370 740)
************
741
(TEST-COMPLETED 224 741)
************
742
(TEST-COMPLETED 371 742)
************
743
(TEST-COMPLETED 230 743)
************
744
(TEST-COMPLETED 372 744)
************
745
(TEST-COMPLETED 236 745)
************
746
(TEST-COMPLETED 373 746)
************
747
(TEST-COMPLETED 242 747)
************
748
(TEST-COMPLETED 374 748)
************
749
(TEST-COMPLETED 248 749)
************
750
(TEST-COMPLETED 375 750)
************
751
(TEST-COMPLETED 254 751)
************
752
(TEST-COMPLETED 376 752)
************
753
```

```
(TEST-COMPLETED 260 753)
* * * * * * * * * * * *
754
(TEST-COMPLETED 377 754)
* * * * * * * * * * * *
755
(TEST-COMPLETED 266 755)
* * * * * * * * * * * *
756
(TEST-COMPLETED 378 756)
* * * * * * * * * * * *
757
(TEST-COMPLETED 272 757)
* * * * * * * * * * * *
758
(TEST-COMPLETED 379 758)
* * * * * * * * * * * *
759
(TEST-COMPLETED 278 759)
* * * * * * * * * * * *
760
(TEST-COMPLETED 380 760)
* * * * * * * * * * * *
761
(TEST-COMPLETED 284 761)
* * * * * * * * * * * *
762
(TEST-COMPLETED 381 762)
* * * * * * * * * * * *
763
(TEST-COMPLETED 290 763)
* * * * * * * * * * * *
764
(TEST-COMPLETED 382 764)
* * * * * * * * * * * *
765
(TEST-COMPLETED 296 765)
* * * * * * * * * * * *
766
(TEST-COMPLETED 383 766)
* * * * * * * * * * * *
767
(TEST-COMPLETED 302 767)
* * * * * * * * * * * *
768
(TEST-COMPLETED 384 768)
* * * * * * * * * * * *
769
(TEST-COMPLETED 308 769)
* * * * * * * * * * * *
770
(TEST-COMPLETED 385 770)
* * * * * * * * * * * *
771
(TEST-COMPLETED 314 771)
* * * * * * * * * * * *
772
```

```
(TEST-COMPLETED 386 772)
************
773
(TEST-COMPLETED 320 773)
************
774
(TEST-COMPLETED 387 774)
************
775
(TEST-COMPLETED 326 775)
************
776
(TEST-COMPLETED 388 776)
************
777
(TEST-COMPLETED 332 777)
************
778
(TEST-COMPLETED 389 778)
************
779
(TEST-COMPLETED 338 779)
************
780
(TEST-COMPLETED 390 780)
************
781
(TEST-COMPLETED 344 781)
************
782
(TEST-COMPLETED 391 782)
************
783
(TEST-COMPLETED 350 783)
************
784
(TEST-COMPLETED 392 784)
************
785
(TEST-COMPLETED 356 785)
************
786
(TEST-COMPLETED 393 786)
************
787
(TEST-COMPLETED 362 787)
************
788
(TEST-COMPLETED 394 788)
************
789
(TEST-COMPLETED 368 789)
************
790
(TEST-COMPLETED 395 790)
************
791
```

```
(TEST-COMPLETED 374 791)
************
792
(TEST-COMPLETED 396 792)
************
793
(TEST-COMPLETED 380 793)
************
794
(TEST-COMPLETED 397 794)
************
795
(TEST-COMPLETED 386 795)
************
796
(TEST-COMPLETED 398 796)
************
797
(TEST-COMPLETED 392 797)
************
798
(TEST-COMPLETED 399 798)
************
799
(TEST-COMPLETED 398 799)
************
800
(TEST-COMPLETED 400 800)
************
801
(TEST-COMPLETED 404 801)
************
802
(TEST-COMPLETED 401 802)
************
803
(TEST-COMPLETED 410 803)
************
804
(TEST-COMPLETED 402 804)
************
805
(TEST-COMPLETED 416 805)
************
806
(TEST-COMPLETED 403 806)
************
807
(TEST-COMPLETED 422 807)
************
808
(TEST-COMPLETED 404 808)
************
809
(TEST-COMPLETED 428 809)
************
810
```

```
(TEST-COMPLETED 405 810)
************
811
(TEST-COMPLETED 434 811)
************
812
(TEST-COMPLETED 406 812)
************
813
(TEST-COMPLETED 440 813)
************
814
(TEST-COMPLETED 407 814)
************
815
(TEST-COMPLETED 446 815)
************
816
(TEST-COMPLETED 408 816)
************
817
(TEST-COMPLETED 452 817)
************
818
(TEST-COMPLETED 409 818)
************
819
(TEST-COMPLETED 458 819)
************
820
(TEST-COMPLETED 410 820)
************
821
(TEST-COMPLETED 464 821)
************
822
(TEST-COMPLETED 411 822)
************
823
(TEST-COMPLETED 470 823)
************
824
(TEST-COMPLETED 412 824)
************
825
(TEST-COMPLETED 476 825)
************
826
(TEST-COMPLETED 413 826)
************
827
(TEST-COMPLETED 482 827)
************
828
(TEST-COMPLETED 414 828)
************
829
```

```
(TEST-COMPLETED 488 829)
************
830
(TEST-COMPLETED 415 830)
************
831
(TEST-COMPLETED 494 831)
************
832
(TEST-COMPLETED 416 832)
************
833
(TEST-COMPLETED 500 833)
************
834
(TEST-COMPLETED 417 834)
************
835
(TEST-COMPLETED 506 835)
************
836
(TEST-COMPLETED 418 836)
************
837
(TEST-COMPLETED 512 837)
************
838
(TEST-COMPLETED 419 838)
************
839
(TEST-COMPLETED 518 839)
************
840
(TEST-COMPLETED 420 840)
************
841
(TEST-COMPLETED 524 841)
************
842
(TEST-COMPLETED 421 842)
************
843
(TEST-COMPLETED 530 843)
************
844
(TEST-COMPLETED 422 844)
************
845
(TEST-COMPLETED 536 845)
************
846
(TEST-COMPLETED 423 846)
************
847
(TEST-COMPLETED 542 847)
************
848
```

```
(TEST-COMPLETED 424 848)
************
849
(TEST-COMPLETED 548 849)
************
850
(TEST-COMPLETED 425 850)
************
851
(TEST-COMPLETED 554 851)
************
852
(TEST-COMPLETED 426 852)
************
853
(TEST-COMPLETED 560 853)
************
854
(TEST-COMPLETED 427 854)
************
855
(TEST-COMPLETED 566 855)
************
856
(TEST-COMPLETED 428 856)
************
857
(TEST-COMPLETED 572 857)
************
858
(TEST-COMPLETED 429 858)
************
859
(TEST-COMPLETED 578 859)
************
860
(TEST-COMPLETED 430 860)
************
861
(TEST-COMPLETED 584 861)
************
862
(TEST-COMPLETED 431 862)
************
863
(TEST-COMPLETED 590 863)
************
864
(TEST-COMPLETED 432 864)
************
865
(TEST-COMPLETED 596 865)
************
866
(TEST-COMPLETED 433 866)
************
867
```

```
(TEST-COMPLETED 602 867)
************
868
(TEST-COMPLETED 434 868)
************
869
(TEST-COMPLETED 608 869)
************
870
(TEST-COMPLETED 435 870)
************
871
(TEST-COMPLETED 614 871)
************
872
(TEST-COMPLETED 436 872)
************
873
(TEST-COMPLETED 620 873)
************
874
(TEST-COMPLETED 437 874)
************
875
(TEST-COMPLETED 626 875)
************
876
(TEST-COMPLETED 438 876)
************
877
(TEST-COMPLETED 632 877)
************
878
(TEST-COMPLETED 439 878)
************
879
(TEST-COMPLETED 638 879)
************
880
(TEST-COMPLETED 440 880)
************
881
(TEST-COMPLETED 644 881)
************
882
(TEST-COMPLETED 441 882)
************
883
(TEST-COMPLETED 650 883)
************
884
(TEST-COMPLETED 442 884)
************
885
(TEST-COMPLETED 656 885)
************
886
```

```
(TEST-COMPLETED 443 886)
************
887
(TEST-COMPLETED 662 887)
************
888
(TEST-COMPLETED 444 888)
************
889
(TEST-COMPLETED 668 889)
************
890
(TEST-COMPLETED 445 890)
************
891
(TEST-COMPLETED 674 891)
************
892
(TEST-COMPLETED 446 892)
************
893
(TEST-COMPLETED 680 893)
************
894
(TEST-COMPLETED 447 894)
************
895
(TEST-COMPLETED 686 895)
************
896
(TEST-COMPLETED 448 896)
************
897
(TEST-COMPLETED 692 897)
************
898
(TEST-COMPLETED 449 898)
************
899
(TEST-COMPLETED 698 899)
************
900
(TEST-COMPLETED 450 900)
************
901
(TEST-COMPLETED 704 901)
************
902
(TEST-COMPLETED 451 902)
************
903
(TEST-COMPLETED 710 903)
************
904
(TEST-COMPLETED 452 904)
************
905
```

```
(TEST-COMPLETED 716 905)
************
906
(TEST-COMPLETED 453 906)
************
907
(TEST-COMPLETED 722 907)
************
908
(TEST-COMPLETED 454 908)
************
909
(TEST-COMPLETED 728 909)
************
910
(TEST-COMPLETED 455 910)
************
911
(TEST-COMPLETED 734 911)
************
912
(TEST-COMPLETED 456 912)
************
913
(TEST-COMPLETED 740 913)
************
914
(TEST-COMPLETED 457 914)
************
915
(TEST-COMPLETED 746 915)
************
916
(TEST-COMPLETED 458 916)
************
917
(TEST-COMPLETED 752 917)
************
918
(TEST-COMPLETED 459 918)
************
919
(TEST-COMPLETED 758 919)
************
920
(TEST-COMPLETED 460 920)
************
921
(TEST-COMPLETED 764 921)
************
922
(TEST-COMPLETED 461 922)
************
923
(TEST-COMPLETED 770 923)
************
924
```

```
(TEST-COMPLETED 462 924)
************
925
(TEST-COMPLETED 776 925)
************
926
(TEST-COMPLETED 463 926)
************
927
(TEST-COMPLETED 782 927)
************
928
(TEST-COMPLETED 464 928)
************
929
(TEST-COMPLETED 788 929)
************
930
(TEST-COMPLETED 465 930)
************
931
(TEST-COMPLETED 794 931)
************
932
(TEST-COMPLETED 466 932)
************
933
(TEST-COMPLETED 800 933)
************
934
(TEST-COMPLETED 467 934)
************
935
(TEST-COMPLETED 806 935)
************
936
(TEST-COMPLETED 468 936)
************
937
(TEST-COMPLETED 812 937)
************
938
(TEST-COMPLETED 469 938)
************
939
(TEST-COMPLETED 818 939)
************
940
(TEST-COMPLETED 470 940)
************
941
(TEST-COMPLETED 824 941)
************
942
(TEST-COMPLETED 471 942)
************
943
```

```
(TEST-COMPLETED 830 943)
************
944
(TEST-COMPLETED 472 944)
************
945
(TEST-COMPLETED 836 945)
************
946
(TEST-COMPLETED 473 946)
************
947
(TEST-COMPLETED 842 947)
************
948
(TEST-COMPLETED 474 948)
************
949
(TEST-COMPLETED 848 949)
************
950
(TEST-COMPLETED 475 950)
************
951
(TEST-COMPLETED 854 951)
************
952
(TEST-COMPLETED 476 952)
************
953
(EVEN 860 953)
(BRANCHING 430)
(TEST-COMPLETED 215 430)
(BRANCHING 930)
(TEST-COMPLETED 465 930)
************
954
(TEST-COMPLETED 477 954)
************
955
(EVEN 866 955)
(BRANCHING 433)
(TEST-COMPLETED 300 433)
(BRANCHING 933)
(TEST-COMPLETED 800 933)
************
956
(TEST-COMPLETED 478 956)
************
957
(EVEN 872 957)
(BRANCHING 436)
(TEST-COMPLETED 218 436)
(BRANCHING 936)
(TEST-COMPLETED 468 936)
************
958
```

```
(TEST-COMPLETED 479 958)
************
959
(EVEN 878 959)
(BRANCHING 439)
(TEST-COMPLETED 318 439)
(BRANCHING 939)
(TEST-COMPLETED 818 939)
************
960
(TEST-COMPLETED 480 960)
************
961
(EVEN 884 961)
(BRANCHING 442)
(TEST-COMPLETED 221 442)
(BRANCHING 942)
(TEST-COMPLETED 471 942)
************
962
(TEST-COMPLETED 481 962)
************
963
(EVEN 890 963)
(BRANCHING 445)
(TEST-COMPLETED 336 445)
(BRANCHING 945)
(TEST-COMPLETED 836 945)
************
964
(TEST-COMPLETED 482 964)
************
965
(EVEN 896 965)
(BRANCHING 448)
(TEST-COMPLETED 224 448)
(BRANCHING 948)
(TEST-COMPLETED 474 948)
************
966
(TEST-COMPLETED 483 966)
************
967
(EVEN 902 967)
(BRANCHING 451)
(TEST-COMPLETED 354 451)
(BRANCHING 951)
(TEST-COMPLETED 854 951)
************
968
(TEST-COMPLETED 484 968)
************
969
(EVEN 908 969)
(BRANCHING 454)
(TEST-COMPLETED 227 454)
(BRANCHING 954)
```

```
(TEST-COMPLETED 477 954)
************
970
(TEST-COMPLETED 485 970)
************
971
(EVEN 914 971)
(BRANCHING 457)
(TEST-COMPLETED 372 457)
(BRANCHING 957)
(EVEN 872 957)
(BRANCHING 436)
(TEST-COMPLETED 218 436)
(BRANCHING 936)
(TEST-COMPLETED 468 936)
************
972
(TEST-COMPLETED 486 972)
************
973
(EVEN 920 973)
(BRANCHING 460)
(TEST-COMPLETED 230 460)
(BRANCHING 960)
(TEST-COMPLETED 480 960)
************
974
(TEST-COMPLETED 487 974)
************
975
(EVEN 926 975)
(BRANCHING 463)
(TEST-COMPLETED 390 463)
(BRANCHING 963)
(EVEN 890 963)
(BRANCHING 445)
(TEST-COMPLETED 336 445)
(BRANCHING 945)
(TEST-COMPLETED 836 945)
************
976
(TEST-COMPLETED 488 976)
************
977
(EVEN 932 977)
(BRANCHING 466)
(TEST-COMPLETED 233 466)
(BRANCHING 966)
(TEST-COMPLETED 483 966)
************
978
(TEST-COMPLETED 489 978)
************
979
(EVEN 938 979)
(BRANCHING 469)
(TEST-COMPLETED 408 469)
```

```
(BRANCHING 969)
(EVEN 908 969)
(BRANCHING 454)
(TEST-COMPLETED 227 454)
(BRANCHING 954)
(TEST-COMPLETED 477 954)
************
980
(TEST-COMPLETED 490 980)
************
981
(EVEN 944 981)
(BRANCHING 472)
(TEST-COMPLETED 236 472)
(BRANCHING 972)
(TEST-COMPLETED 486 972)
************
982
(TEST-COMPLETED 491 982)
************
983
(EVEN 950 983)
(BRANCHING 475)
(TEST-COMPLETED 426 475)
(BRANCHING 975)
(EVEN 926 975)
(BRANCHING 463)
(TEST-COMPLETED 390 463)
(BRANCHING 963)
(EVEN 890 963)
(BRANCHING 445)
(TEST-COMPLETED 336 445)
(BRANCHING 945)
(TEST-COMPLETED 836 945)
************
984
(TEST-COMPLETED 492 984)
************
985
(EVEN 956 985)
(BRANCHING 478)
(TEST-COMPLETED 239 478)
(BRANCHING 978)
(TEST-COMPLETED 489 978)
************
986
(TEST-COMPLETED 493 986)
************
987
(EVEN 962 987)
(BRANCHING 481)
(EVEN 444 481)
(BRANCHING 222)
(TEST-COMPLETED 111 222)
(BRANCHING 722)
(TEST-COMPLETED 361 722)
(BRANCHING 981)
```

```
(EVEN 944 981)
(BRANCHING 472)
(TEST-COMPLETED 236 472)
(BRANCHING 972)
(TEST-COMPLETED 486 972)
************
988
(TEST-COMPLETED 494 988)
************
989
(EVEN 968 989)
(BRANCHING 484)
(TEST-COMPLETED 242 484)
(BRANCHING 984)
(TEST-COMPLETED 492 984)
************
990
(TEST-COMPLETED 495 990)
************
991
(EVEN 974 991)
(BRANCHING 487)
(EVEN 462 487)
(BRANCHING 231)
(EVEN 694 231)
(BRANCHING 347)
(TEST-COMPLETED 42 347)
(BRANCHING 847)
(TEST-COMPLETED 542 847)
(BRANCHING 731)
(TEST-COMPLETED 194 731)
(BRANCHING 987)
(EVEN 962 987)
(BRANCHING 481)
(EVEN 444 481)
(BRANCHING 222)
(TEST-COMPLETED 111 222)
(BRANCHING 722)
(TEST-COMPLETED 361 722)
(BRANCHING 981)
(EVEN 944 981)
(BRANCHING 472)
(TEST-COMPLETED 236 472)
(BRANCHING 972)
(TEST-COMPLETED 486 972)
************
992
(TEST-COMPLETED 496 992)
************
993
(EVEN 980 993)
(BRANCHING 490)
(TEST-COMPLETED 245 490)
(BRANCHING 990)
(TEST-COMPLETED 495 990)
************
994
```

```
(TEST-COMPLETED 497 994)
************
995
(EVEN 986 995)
(BRANCHING 493)
(EVEN 480 493)
(BRANCHING 240)
(TEST-COMPLETED 120 240)
(BRANCHING 740)
(TEST-COMPLETED 370 740)
(BRANCHING 993)
(EVEN 980 993)
(BRANCHING 490)
(TEST-COMPLETED 245 490)
(BRANCHING 990)
(TEST-COMPLETED 495 990)
************
996
(TEST-COMPLETED 498 996)
************
997
(EVEN 992 997)
(BRANCHING 496)
(TEST-COMPLETED 248 496)
(BRANCHING 996)
(TEST-COMPLETED 498 996)
************
998
(TEST-COMPLETED 499 998)
************
```

```
(EVEN 998 999)
(BRANCHING 499)
(EVEN 498 999)
(BRANCHING 249)
(TEST-COMPLETED 748 999)
(BRANCHING 749)
(TEST-COMPLETED 248 999)
(BRANCHING 999)
(EVEN 998 999)
(BRANCHING 499)
(TEST-COMPLETED 498 999)
(BRANCHING 999)
(TEST-COMPLETED 998 999)
```